Procreative Justice

This book explores how considerations of justice apply to procreative decision-making.

Despite its immense personal significance, procreation is an inherently other-regarding endeavor. By its very nature, the decision to procreate is the decision to bring into existence another morally considerable being, one who will be exposed to the full range of harms, benefits, and risks that accompany a typical human life, and one who cannot by its nature ever consent to being born. Moreover, when this decision is undertaken in a community of persons, it is also a decision to affect the lives of others in a host of profound if often underappreciated ways, from its effects on population size and environmental sustainability to its consequences for a community's distribution of resources. In many ways, of course, these interests coincide: adults need children for their parenting projects, societies need citizens for the maintenance of their institutions, and children themselves are often happy to have been brought into existence. However, as the book demonstrates, the various interests that are implicated by procreative decision-making can also come into conflict as well and in ways that raise basic questions of justice. Through a systematic examination of six of these questions, the author argues that taking adequate stock of the conflicting interests at stake in procreative decision-making leads to a narrower view of the conditions under which it is morally permissible to procreate and a much more demanding conception of our procreative responsibilities.

Procreative Justice will be of interest to researchers and advanced students working on the morality of procreation and related areas of philosophy, including bioethics, intergenerational ethics, environmental ethics, population ethics, and the ethics of the family.

Erik Magnusson is a political philosopher based at the University of Manitoba, where he serves as a research facilitator in the Faculty of Education, a sessional instructor in the Departments of Political Studies and Philosophy, and an Associate of the Centre for Professional and Applied Ethics.

Routledge Research in Applied Ethics

New Omnivorism and Strict Veganism
Critical Perspectives
Edited by Cheryl Abbate and Christopher Bobier

Human Freedom in the Age of AI
Filippo Santoni de Sio

Procreative Responsibility and Assisted Reproductive Technologies
Davide Battisti

Ethics and Situational Crime Prevention
Thomas Søbirk Petersen

Rethinking Informed Consent in the Big Data Age
Adam J. Andreotta

A Philosophical Case for Ecological Pessimism
Toby Svoboda

Ending Wars Justly
Theory and Applications
Edited by David K. Chan

Procreative Justice
Balancing the Interests of Parents, Children, and Society
Erik Magnusson

For more information about this series, please visit: https://www.routledge.com/Routledge-Research-in-Applied-Ethics/book-series/RRAES

Procreative Justice

Balancing the Interests of Parents, Children, and Society

Erik Magnusson

Routledge
Taylor & Francis Group
NEW YORK AND LONDON

First published 2026
by Routledge
605 Third Avenue, New York, NY 10158

and by Routledge
4 Park Square, Milton Park, Abingdon, Oxon, OX14 4RN

Routledge is an imprint of the Taylor & Francis Group, an informa business

ISBN: 978-0-367-75696-3 (hbk)
ISBN: 978-0-367-75816-5 (pbk)
ISBN: 978-1-003-16410-4 (ebk)

DOI: 10.4324/9781003164104

Typeset in Sabon
by SPi Technologies India Pvt Ltd (Straive)

Contents

Acknowledgments *vii*

Introduction 1

1 The Right to Procreate 11

1.1 Introduction 11
1.2 The Interest Theory of Rights 13
1.3 The Nature of the Interest in Procreation 18
1.4 The Parental Interest Justification of the Right to Procreate 24
1.5 The Content and Limits of the Right to Procreate 36

2 The Child's Right to a Decent Start 46

2.1 Introduction 46
2.2 The Non-Identity Problem 48
2.3 A Rights-Based Solution 50
2.4 Objections 54
2.5 The Decent Start Principle 68

3 Procreation and Parental Rights 74

3.1 Introduction 74
3.2 Disambiguating Parental Rights and Obligations 76
3.3 Parental Rights and the Baby Redistribution Problem 77
3.4 The Relationship-Based Gestational Account 82
3.5 Three Problems 86
3.6 The Parental Project View 95

4 **Procreation and Parental Obligations** 107

4.1 *Introduction* 107
4.2 *Theories of Parental Obligation* 109
4.3 *Defending the Causal Account of Parental Obligation* 118
4.4 *Toward Moral Responsibility for Parental Need* 139
4.5 *Parenthood Unpackaged* 140

5 **Distributing the Cost of Procreation** 148

5.1 *Introduction* 148
5.2 *The Responsibility Argument* 150
5.3 *The Public Goods Argument* 153
5.4 *Why Socialize the Benefits of Child-Rearing?* 163
5.5 *Internal and External Limitations on the Pro-Sharing View* 170

6 **Procreation and Adoption** 178

6.1 *Introduction* 178
6.2 *The Rescue-Based Argument for the Duty to Adopt* 180
6.3 *Under What Conditions Do We Have a Duty to Rescue?* 183
6.4 *Can a Moderate Duty to Rescue Generate a Duty to Adopt?* 187
6.5 *Collective Obligations to Facilitate Adoption: The Case of ART* 198

7 **Procreation and the Environment** 207

7.1 *Introduction* 207
7.2 *The Environmental Argument for Limiting Procreation* 209
7.3 *Procreation and Responsibility for Environmental Impact* 211
7.4 *How Many Children Is Too Many?* 217
7.5 *Are Procreation and Consumption Morally Equivalent?* 228

Bibliography 234
Index 245

Acknowledgments

This book has been in development for a long time and has accordingly benefitted from the ideas, advice, and encouragement of many people. The content of this book originated in my doctoral research undertaken at the University of Oxford from 2012 to 2016, where I received invaluable guidance from my thesis supervisor Simon Caney; my various committee members and examiners, including Daniel Butt, Matthew Clayton, Zofia Stemplowska, and Jeremy Waldron; as well as members of the broader philosophical community, particularly in the Oxford Graduate Political Theory Workshop and the Oxford Applied Ethics Discussion Group. Following my time in Oxford, I was very fortunate to hold a post-doctoral research fellowship at Pompeu Fabra University in Barcelona, Spain, from 2016 to 2018, where I had the opportunity to develop my research further under the auspice of a European Research Council project on "Justice and the Family: An Analysis of the Normative Significance of Procreation and Parenting for a Just Society" led by Serena Olsaretti. I benefitted tremendously from Serena's mentorship and supervision, as well as the support of my wonderful colleagues in the Family Justice Research Group, including Paul Bou-Habib, Andrée-Anne Cormier, Anca Gheaus, Riccardo Spotorno, and Isabella Trifan. During the many years that I have worked on this material, I have had fruitful discussions with, and received helpful feedback from, far too many people to mention, but the following come to mind: Gustaf Arrhenius, Nicolás Brando, Paula Casal, Elizabeth Finneron-Burns, William Gildea, Sarah Hannan, Darian Heim, Derek Kornelsen, Steven Lecce, RJ Leland, Julian López-Murcia, Jordan Mansell, Tim Meijers, Seliha Metinsoy, Andres Moles, Tom Parr, Bryan Peeler, Mauro Rossi, Julian Savulescu, Rob Shaver, Liam Shields, Adam Swift, Richard Vernon, and Andrew Williams.

I am grateful to Andrew Weckenmann at Routledge for his initial enthusiasm for this project, as well as to Rosaleah Stammler for being especially patient with me as I hobbled across the finish line. I am also appreciative of two anonymous reviewers who provided helpful feedback on the manuscript.

Some of the content of this book has appeared elsewhere in earlier versions. Parts of Chapter 2 were published previously as "Children's Rights and the Non-Identity Problem," *Canadian Journal of Philosophy*, 49(5) (2019), pp. 580–605; parts of Chapter 3 were published previously as "Can Gestation Ground Parental Rights?" *Social Theory and Practice*, 46(1) (2020), pp. 111–142; parts of Chapter 5 were published previously as "Parental Justice and the Kids Pay View," *Ethical Theory and Moral Practice*, 21(4) (2018), pp. 963–977; and parts of Chapter 6 were published previously as "Do Prospective Parents Have a Duty to Adopt Rather than Procreate?" *Public Affairs Quarterly*, 39(1) (2025), pp. 66–86. I am grateful to the publishers for permission to reproduce some of this material here and to the anonymous reviewers from those journals whose comments helped to improve my arguments.

Research for this project was supported at various stages by the Social Sciences and Humanities Research Council of Canada (Grant #:106646) and the European Research Council (Grant #:648610). I am grateful to these agencies for their financial support.

Finally, I am grateful to members of my family who have supported me along the way in too many ways to mention, including David Magnusson, Jean Merrell, Laura Magnusson, Julio Godoy, Maria Beatriz Verdun, Santiago Godoy, and my beloved writing companion, Bernie the cockapoo. Most of all, however, I am grateful to my procreative partner, Maria Godoy, and to our beautiful progeny, Freya and Theo, who remind me every day of the profound value that is at stake in creating and raising new human beings. This book is dedicated to them.

Introduction

Imagine the following scenario. You and your partner have washed up on the shore of a desert island following a shipwreck, along with a small group of the ship's surviving passengers. It soon becomes clear that you will be stranded there for the foreseeable future—given the remoteness of the island, combined with the ship's faulty communications equipment, the chances of a rescue are slim to none. Fortunately, however, the island is plentiful: there are a variety of fruits, nuts, and root vegetables native to the soil, and an abundance of coconut timber to construct sturdy and comfortable dwellings. After resigning yourselves to your newfound predicament, you and the other passengers quickly get to work, gathering basic resources and building temporary shelters to ensure your immediate survival.

One year later, you and the other passengers have begun settling into life on the island. You have collectively built a small village of permanent dwellings and have even started a cooperative economy based on a division of labor. There are fishers, gardeners, fruit-pickers, and builders, as well as a variety of skilled artisans and tradespeople, all of whom play an important role in contributing to the communal stock of resources. You and your partner find yourselves surprisingly satisfied with life on the island: the weather is great, you have plenty of time for leisure, and you find genuine meaning in the contributions you make to the wider community. Nevertheless, something is missing. Prior to the shipwreck, you and your partner were planning to conceive your first child, and you have not come to peace with the fact that you have not been able to do so—after all, parenting a biological child is something that is deeply important to you, so much so that you will not consider your life to be a success until you have at least tried to bring a child into existence. After talking it over with your partner, you both decide to pursue your ambition in the face of the nonideal circumstances in which you find yourselves. You will initiate a pregnancy and bring into existence the island's first natural-born resident.

DOI: 10.4324/9781003164104-1

After sharing the news with the other islanders, however, you are surprised to find that your announcement is met with nervous glances and half-hearted congratulations. You ask them what is wrong. "We're happy for you," one of them says,

> but we're just not sure you've thought this through. For one thing, you would be bringing your child into a very precarious situation. Things have been going well on the island so far, but who knows what the future holds? It doesn't seem fair to bring a child into existence when you cannot ensure they will have a decent standard of living. Also, you have to keep in mind that the carrying capacity of the island is limited. We have enough resources to sustain us now, but if we all created additional children, we might be consuming the island's resources at an unsustainable rate. In any event, the resource share of each resident may have to be diminished to accommodate new arrivals, and this seems unfair to those of us who have no interest in parenting. Finally, if you really want to become a parent, there are already numerous children on the island who were orphaned in the shipwreck. Other things being equal, doesn't it make more sense to meet their parenting needs before creating additional needs to be met?

After a moment of stunned silence, you try to explain that, from your perspective, other things are *not* equal: having a biological child is not merely a personal preference, but rather a deeply embedded interest, one you feel you must satisfy in order to give meaning and purpose to your life. "We understand that," another islander says,

> but we all live together in a community, and your interests are not the only interests that count. No one is going to prevent you from having a child if that's what you decide to do. All that we ask is that, before you do, you seriously consider how others will be impacted by your decision.

In the real world, of course, people do not typically behave like your interlocutors on the island. Rather than being called upon to justify their decisions to have children, procreators are more likely to be celebrated for those decisions: they are met with congratulations, they are showered with gifts, and they are often afforded a social status that is not typically available to those who remain childless.[1] However, the island is a microcosm for the world in which we live, and the conditions that make procreation morally troublesome on the island also exist in the real world as well, albeit on a grander scale. Despite its immense personal significance, procreation is an inherently *other-regarding* endeavor. By its very nature, the decision to procreate is the decision to bring into existence another morally considerable being, one who will be exposed to the full range of harms, benefits, and risks that accompany a typical human life, and one who cannot by its nature ever consent to being born. When this decision is undertaken in a community of persons, it is also a decision to affect the lives of others in a host of profound if often underappreciated ways, from its effects on population size and environmental sustainability to its consequences for a community's distribution of resources. In many ways, of course, these interests coincide: adults need children for their parenting projects, societies need citizens for the maintenance of their institutions, and children themselves are often happy to have been brought into existence. As the island scenario reminds us, however, the various interests that are implicated by procreative decision-making can also come into conflict as well, and in ways that raise basic questions of justice.

This book explores what justice requires in the area of procreative decision-making. The *question* of procreative justice is the question of how to fairly balance the competing interests that are at stake in the creation of new people. A *theory* of procreative justice is an answer to this question: it offers a specification of the rights and obligations of individuals *qua* procreators and a set of prescriptions for how the benefits and burdens of procreation ought to be distributed throughout society (and, in some respects, across societies). The purpose of this book is to develop a comprehensive theory of procreative justice that can be used to provide normative guidance in a wide range of ethical controversies involving procreation, from the conditions under which it is morally permissible to bring children into existence, to the allocation of parental rights and obligations in custody disputes over newborn children, to the social distribution of child-rearing costs more generally. It does so by addressing six interrelated questions pertaining to our procreative rights and obligations:

1 Is there a moral right to procreate? If so, what is the nature of the interest that grounds it, and what kinds of duties does it impose on others?
2 Can procreation ever be wrongful in light of its effects on the resultant child? What role (if any) do prospective children's interests play in limiting the conditions under which it is morally permissible to procreate?
3 What is the nature of the moral relationship between procreation and parenthood? Does the act of creating a child generate a special moral right and/or obligation to parent that child?
4 How should the significant costs of procreation and parenting be distributed throughout society? Should parents be responsible for paying the full cost of their children's upbringing, or should child-rearing costs be shared in some way between parents and non-parents?

5 How do the claims of orphaned children impact the morality of procreation? Is it morally permissible to bring new children into existence when there are already existing children in need of parental care?
6 How do our climate-related duties impact the morality of procreation? Given the significant carbon impact of procreation, are we morally obligated to limit the number of children we create?

These questions are diverse in their focus: some have to do with the rights and interests of procreators; some have to do with the rights and interests of the children they bring into existence; and some have to do with the rights and interests of the wide range of third parties who may be impacted by the addition of new members into the community, including taxpayers, orphaned children, and members of present and future generations who are vulnerable to the environmental impact of population growth. However, they are all closely interconnected, and one of the objectives of this book is to show how they are best addressed in conjunction with one another rather than being treated in isolation. For instance, the position we take on the grounds and content of the right to procreate will strongly influence the position we take on debates surrounding the duty to adopt or environmental obligations to have fewer children. Similarly, how we appraise different theories of the origins of parental obligations will vary considerably depending on the position we take regarding the social distribution of child-rearing costs more generally. In covering this wide range of questions and demonstrating how they are interdependent, this book aims to bring together what have sometimes been siloed debates about the morality of procreation and, in so doing, provide a more comprehensive and integrated theory of procreative justice that has previously been offered in the existing literature.

Chapter 1 begins by considering the nature of the interest in procreation, and asking what rights (if any) that interest can support. In the existing literature surrounding the morality of procreation, it is often assumed that adults have a moral right to procreate and that this right supports claims to non-interference or even positive assistance in a wide range of procreative decisions, including decisions about whether, when, with whom, and how often to procreate, and even what kinds of children to create through the application of selective reproductive technologies. However, while much attention has been paid to analyzing the importance of the right to procreate and the range of decisions it might be called on to support, relatively little attention has been paid to examining its philosophical foundations, including the nature of the interest that grounds it or the content of the duties it imposes on others. In Chapter 1, I draw on the interest theory of rights to shed light on these issues and, in so doing, advance two main claims. The first claim is that the right to procreate is best understood as deriving from a more fundamental right to parent, which is itself grounded in the strong interest that adults have in developing intimate fiduciary relationships with children. In short, adults have a morally significant interest in enjoying a certain type of relationship with children, and procreation offers the primary (though not always the only) means by which such a relationship can be justly initiated. The second claim is that clarifying the interest that grounds the right to procreate has important implications for its content, including the nature of the duties it imposes on others, the circumstances in which it may be appropriately exercised or claimed, and the relative weight that it carries in conflicts with competing moral considerations. In this sense, when understood properly, the right to procreate protects a much narrower range of procreative decisions than is commonly assumed, and some of the decisions currently defended under its name—including the decision to procreate without parenting or the decision to create as many children as one chooses—are in fact outside of its scope.

Having explored the grounds and content of the right to procreate in Chapter 1, I turn in Chapter 2 to consider whether the exercise of this right can ever be constrained in light of the interests of the children who may be created as a result of its exercise. Given the profound impact of procreation on the welfare and life prospects of the children who result from it, it is reasonable to think that the right to procreate ought to be constrained in light of their interests, perhaps only permitting the creation of children who can reasonably be expected to live minimally decent lives. This is consistent with ordinary understandings of the scope and limits of other parenthood-related rights, for example, whose acquisition and maintenance is normally conditional on promoting the child's interests to some minimum degree. However, there is a philosophical problem in the case of procreation that complicates this reasoning and threatens to limit the extent to which children's interests can be incorporated into a theory of procreative justice. When a set of parents fail to promote their existing child's interests, there is an identifiable person who is made worse off by that failure: their child's life goes worse than it otherwise would have gone had they fulfilled their responsibilities and promoted the child's interests. However, the same is not true in the case of procreation. When a set of procreators bring their child into disadvantaged conditions, the child is not usually made any worse off than she otherwise would have been, as their very existence depended on that procreative act taking place when and how it did. This, in turn, raises an important and notoriously difficult question: how can an act be contrary to a person's interests when it is also a necessary condition of their existence?

Chapter 2 addresses this question by examining the non-identity problem and considering its implications for the morality of procreation. I argue against the terms of that problem that even if a child cannot be made comparatively worse off by being born into disadvantaged conditions, they may nevertheless be wronged if they are born into conditions that fall below a minimally decent welfare threshold to which all existing children are entitled. The wrong in question, I argue, can be understood in terms of violating the child's *right to a decent start*, a right that functions to limit the scope of their would-be parents' right to procreate.

Once a child is permissibly brought into existence, they have a legitimate claim to adequate parental care for the duration of their childhood. This raises two important, though ultimately separable questions: (1) who is morally entitled to act as the child's parent? and (2) who is morally responsible for acting as the child's parent? Chapter 3 addresses the first question by considering whether causing a child to exist generates a special moral right to parent that child. In law and common-sense morality, it is widely accepted that adults who meet a minimum threshold of parental competency have a presumptive right to parent their biological children. The legitimacy of this right is also assumed in the account of the right to procreate developed in Chapter 1, according to which the right to procreate is grounded in a strong interest that adults have in developing close intimate relationships with children. However, the philosophical justification of this right is far from clear, for while existing accounts of parental rights explain why it is legitimate for adults to exercise authority over children in general, they do not on their own explain how adults come to acquire that authority over specific children. What is needed, then, is a supplementary explanation of why individuals who have a right to parent in general also have a special right to enter into parental relationships with their biological children.

In Chapter 3, I explore this issue by critically examining the most prominent explanation in the recent literature—the *relationship-based gestational account*—according to which the moral right to parent one's biological child is grounded in an intimate relationship that develops between babies and their birth parents during the process of gestation. I argue that this type of view faces three serious problems that ultimately discredit it as a tenable account: one relating to its *explanatory* function, or its ability to explain why adults who have a right to parent in general also have a special right to parent their biological children; one relating to its *adjudicatory* function, or its ability to resolve cases in which there are multiple conflicting claims over who is entitled to parent a newborn child; and one relating to its *normative* implications in an important series of test cases. I then show how an alternative autonomy-based approach toward the right to parent one's biological children can avoid these problems. According to the *parental project view*, the right to parent one's biological children is not rooted in any specific feature of the biological relationship, but rather in a more general right to continue one's justice-respecting projects. This right often protects the rights of gestators to parent the children they have gestated, but not necessarily, and not for reasons that make any essential reference to biological connectedness. If sound, this alternative account has important theoretical and practical implications, suggesting among other things that the dominant approach toward surrogacy dispute resolution is currently misguided.

Chapter 4 shifts the focus from parental rights to parental obligations and asks whether the act of creating a child also generates a special obligation to parent that child. The nature of this question is importantly distinct from the question posed in the previous chapter. Whereas a theory of parental rights is required in order to resolve cases in which there are multiple conflicting claims over who is entitled to parent a newborn child (such as surrogacy disputes), a theory of parental obligation is required in order to resolve cases in which there is disagreement between different parties about who is responsible for parenting a newborn child (such as cases of unintentional fathers following accidental pregnancies). The existing literature on the morality of parenthood offers three main theories about how parental obligations are acquired, which are normally presented as rivals: the *genetic* account, which holds that one acquires an obligation to parent a child by virtue of supplying the reproductive gametes from which the child developed; the *voluntarist* account, which holds that one acquires an obligation to parent a child by consenting to do so; and the *causal* account, which holds that one acquires an obligation to parent a child by standing in the relevant causal relationship to that child's existence and hence need for parental care. In Chapter 4, I show that the attractive features of each theory can be captured under a singular version of the causal account that assigns parental obligations to those who are morally responsible for a child's need for parental care. On this view, which I term the *moral responsibility for parental need* account, being morally responsible for a child's existence is sufficient to generate an inalienable obligation to ensure that the child is adequately parented for the duration of their childhood, which often but not always requires parenting the child oneself.

Having identified the conditions under which it is permissible to bring a child into existence, as well as the considerations that determine who is morally entitled and obligated to parent a child once they are born, I turn in Chapter 5 to consider how the significant costs of procreation and parenthood ought to be distributed throughout society. I begin by considering the view, prominent among responsibility-sensitive egalitarians, that the cost of creating and raising children should be borne primarily, if not entirely, by the individuals who choose to bring them into existence. I then consider the most prominent pro-sharing response to this view, according to which non-parents are obligated to share in the cost of child-rearing in virtue of their status as the future beneficiaries of parents' child-rearing labor, mainly (but not exclusively) through their enjoyment of social security benefits funded by the next generation of citizens. After outlining an attractive iteration of this latter view, I then explain how it is conditional on two further considerations. First, it is conditional on the success of independent arguments for redistributive social welfare institutions that translate child-rearing labor into benefits for others; and second, it is conditional on others actually benefiting from those institutions, a consideration that may vary significantly according to both the rate of procreation and how we conceive of the broader demands of global and intergenerational justice. I provide a contractualist argument that works to satisfy the first condition and attempt to specify the conditions under which the second condition might be satisfied by elaborating the various considerations that work to influence whether procreation and parenting creates net positive externalities.

Following my discussion of the social distribution of child-rearing costs in Chapter 5, I turn in Chapter 6 to consider whether it is morally permissible to bring new children into existence for the sake of parenting when there are already millions of existing children in need of parental care. Several philosophers have answered negatively, arguing that there is a *pro tanto* duty to adopt rather than procreate as a means of fulfilling one's interests in parenting. The most prominent argument for this view in the existing literature is the *rescue-based argument*, which derives an individual duty to adopt rather than procreate from a more general duty to rescue or assist those in need. In Chapter 6, I critically examine the rescue-based argument and explain why it fails. First, I argue that we do not necessarily have a duty to rescue in cases that resemble the global orphan crisis, where one's intervention is merely sufficient to prevent serious harm to a potential victim. Second, I argue that even if we had such a duty, it would not necessarily generate a duty to adopt rather than procreate given the significant financial, emotional, and agency-related costs of adoption, particularly in current, non-ideal conditions. The upshot of these arguments is that the rescue-based argument can only generate a duty to adopt rather than procreate with respect to a relatively small constituency of orphaned children, many of whom are likely to be adopted anyway by willing volunteers. In the vast majority of cases, then, the rescue-based argument does not entail that it is wrong to create new children when there are already existing children in need of parental care.

Finally, in Chapter 7, I explore the environmental impact of procreation and population growth by considering sustainability-based obligations to have fewer children. In recent years, many philosophers have defended the *environmental argument for limiting procreation*, which holds that if it is wrong to consume resources or generate environmental impact beyond a certain limit, then, other things being equal, it is also wrong to create however many children it would take to exceed that limit. In Chapter 7, I critically examine this argument with a view to understanding how our climate-related duties impact the morality of procreation. While I do not disagree that a concern for the environment gives us significant moral reasons to limit the number of children we create—particularly if, as I argue in Chapter 1, the core interest in parenting can be satisfied with the creation of just one child—I argue that proponents of the environmental argument for limiting procreation have failed to demonstrate that we are under a moral duty to do so, or that we necessarily act wrongly by creating some particular number of children. I focus my discussion on three key limitations of this argument. First, I argue that many versions rely on implausible and unworkable conceptions of how much of a child's consumption is legitimately attributable to their procreative parents, leading them to overestimate the ecological impact of procreation. Second, I argue that all versions fail to defend specific and independently justified accounts of the limits of permissible consumption and have therefore failed to demonstrate that the creation of some particular number of children would cause a procreative parent to exceed that limit. Finally, I argue that they fail to acknowledge important qualitative differences between procreation and other forms of consumption that justify treating them differently from a moral point of view.

Before proceeding to these arguments, I would like to clarify one important assumption of this book, and in so doing, point out one glaring omission. Rather than evaluating the moral permissibility of procreation *per se*, I assume throughout this book that it is presumptively permissible to create children and then seek to determine the conditions under which it might not be, e.g., if the resultant child will be born into less than minimally decent conditions or if the creation of additional children would cause one to exceed one's individual carbon budget. However, the assumption that procreation is presumptively permissible has notably been challenged by anti-natalist philosophers who argue that many of the principles that tend to serve as fixed points in our moral reasoning, including principles pertaining to harm, consent, and the imposition of risk, actually support the judgment that it is all-things-considered impermissible to bring children into existence.[2] I am not unimpressed with these arguments and have in fact addressed some of them in detail elsewhere.[3] However, as I do not think it is possible to give them due consideration in a volume of this length, I have chosen to set them aside in favor of questions that arise when we assume that procreation is at least presumptively morally permissible. Further consideration of these important challenges to the morality of procreation will have to await another occasion.

Notes

1 As Christine Overall observes, "In contemporary Western culture, it ironically appears that one needs to have reasons *not* to have children, but no reasons are required to have them ... The choice to procreate is not regarded as needing any thought or justification." See *Why Have Children? The Ethical Debate* (Cambridge, MA: MIT Press, 2012), 2.

2 For prominent defenses of anti-natalism, see, for example, David Benatar. *Better Never to Have Been: The Harm of Coming into Existence* (Oxford: Oxford University Press, 2006); Matti Häyry. "A Rational Cure for Pre-reproductive Stress Syndrome," *Journal of Medical Ethics*, 30(4) (2004), pp. 377–378; and Seana Valentine Shiffrin. "Wrongful Life, Procreative Responsibility, and the Significance of Harm," *Legal Theory*, 5(2) (1999), pp. 117–148.

3 See, for example, Erik Magnusson. "How to Reject Benatar's Asymmetry Argument," *Bioethics*, 33(6) (2019), pp. 674–683 and "On Risk-Based Arguments for Anti-Natalism," *Journal of Value Inquiry*, 56(1) (2022), pp. 101–117.

1 The Right to Procreate

1.1 Introduction

The principal aim of a theory of procreative justice is to provide an account of how the interests of procreators ought to be balanced against the potentially conflicting interests of persons who are affected by their procreative decisions, including the children who may be brought into existence as a result of those decisions as well as the wide range of third parties who may be impacted by the addition of new members into the community. Thus, a natural place to begin in developing such a theory is to consider the nature of the interest in procreation and what (if any) moral rights that interest can support. In the philosophical literature surrounding the morality of procreation, it is often assumed that adults have a moral right to procreate and that this right supports claims to non-interference or even positive assistance in a wide range of procreative decisions, including decisions about whether, when, with whom, and how often to create children, and even what kinds of children to create through the application of selective reproductive technologies.[1] This assumption is also reflected in international human rights legislation, where the right to procreate is routinely presented as a basic human right. For instance, Article 16 of the *Universal Declaration of Human Rights* (1948) states that "Men and women of full age, without any limitation due to race, nationality, or religion, have a fundamental right to found a family," while claiming "The family is the fundamental group unit of society and is entitled to protection by society and the state."[2] The *United Nations Conference on Population and Development* (1994) later clarified and expanded on this right, affirming

> the basic right of all couples and individuals to decide freely and responsibly the number, spacing, and timing of their children, and to have the information and means to do so, and the right to attain the highest standard of reproductive health.[3]

DOI: 10.4324/9781003164104-2

Similar pronouncements have also been made elsewhere, including Article 23(2) of the *International Covenant on Civil and Political Rights* (1976)[4] and Article 12 of the *European Convention on Human Rights* (1953).[5]

However, while considerable attention has been paid to analyzing the importance of the right to procreate and the range of decisions it may be called upon to support, relatively little attention has been paid to examining its normative foundations, including the nature of the interests that grounds it or the content of the duties it imposes on others.[6] In this chapter, I draw on the interest theory of rights to shed light on these issues and, in so doing, advance two main claims. The first claim is that the right to procreate is best understood as deriving from a more fundamental right to parent, which is itself grounded in the strong interest that adults have in developing intimate fiduciary relationships with children. This account of the right to procreate is preferable to existing alternatives, which either (1) are based on interests that are not sufficiently important to ground duties, (2) entail duties that are excessively demanding, (3) do not necessarily entail a set of duties that protect the ability of the right-holder to create and raise their own biologically related child, and hence do not accord with our pre-reflective intuitions about the value of the right to procreate, or (4) cannot explain the special significance that is often associated with restrictions on the right to procreate. The second claim is that clarifying the interest that grounds the right to procreate has important implications for its content, including the nature of the duties it imposes on others, the circumstances in which it may be appropriately exercised or claimed, and the relative weight that it carries in conflicts with competing moral considerations. In this sense, when understood properly, the right to procreate protects a much narrower range of procreative decisions than is commonly assumed, and some of the decisions currently defended under its name—including the decision to procreate without parenting or the decision to create as many children as one chooses—are in fact outside of its scope.

The chapter proceeds in four parts. In Section 1.2, I outline the interest theory of rights and highlight some important features of that theory that are pertinent to the present analysis. In Section 1.3, I consider a number of possible interests involved in procreative decision-making and argue that only one of them—the interest in parenting—is capable of grounding a right to procreate in the way that we normally understand that right. In Section 1.4, I present what I term the *parental interest justification of the right to procreate* and defend it against three important objections. Finally, in Section 1.5, I outline the distinctive content and limits of a right to procreate derived from a more fundamental right to parent.

1.2 The Interest Theory of Rights

Before evaluating the grounds and content of the right to procreate, it is first necessary to say something about what rights are and how they are typically analyzed on the interest theory of rights, which I assume provides the most plausible account of the nature and function of rights.[7] To have a right in the sense I am concerned with here is to have a moral claim against others to the performance of some duty, including both negative duties of non-interference and positive duties to the provision of goods or services. On Wesley Hohfeld's famous taxonomy, these are known as claim-rights, or "rights in the strictest sense."[8] Claim-rights can be contrasted with liberty-rights, which simply refer to the absence of duties. For instance, I have a claim-right to Θ insofar as others are under a duty not to prevent me from Θ or to positively assist me to Θ, whereas I have a liberty-right to Θ insofar as I am under no duty not to Θ. The right to procreate includes both a claim and a liberty component, as it correlates with duties in others and requires that its holder is under no duty not to procreate correlative to someone else's claim. In this chapter, however, I am mainly concerned with the claim component of the right to procreate, including the nature of the interest that grounds it and the content of the duties it imposes on others.

The interest theory is a theory about the function that claim-rights serve for the right-holder, and as such, provides a useful framework for identifying and evaluating different possible rights-claims. According to the interest theory, rights serve primarily to benefit the right-holder by protecting or promoting a certain aspect of their well-being.[9] The most prominent exponent in the contemporary literature is Joseph Raz, who summarizes his view in the following way:

> 'X has a right' if and only if X can have rights, and, other things being equal, an aspect of X's well-being (his interest) is a sufficient reason for holding some other person(s) to be under a duty.[10]

On Raz's account, two main conditions must be satisfied in order for a person to possess a right: first, they must have the capacity to possess rights, in the sense of having the capacity for well-being (call this the *capacity* condition); and second, a particular aspect of their well-being must be sufficiently important to justify holding others to be under a duty (call this the *sufficiency* condition). The capacity condition is important when we are evaluating the rights-claims of marginal members of the moral community, including non-human animals, fetuses, the deceased, and (importantly for later chapters) not-yet-existing persons, though it is not especially important when we are evaluating the rights-claims of existing human adults, who we can safely assume have the capacity for well-being.

The sufficiency condition, however, is central to the present discussion and therefore merits closer examination. What, then, is an interest in the relevant sense, and when does the protection of an interest provide sufficient reason for holding other person(s) to be under a duty?

In the most general terms, a person has an interest in Θ to the extent that Θ contributes in some way to their well-being, where well-being can be understood as a measure of what is non-instrumentally or ultimately good for a person. There are of course many different theories of well-being and hence many different ways of understanding what is in a person's interest, though I will simply assume in what follows that well-being is at least partially understood in objectivist terms, such that a person's interests can be distinct from their preferences or desires. On this view, a person may have no subjective desire to do things like perform exercise or eat a nutritious diet, but this does not prevent us from saying that doing these things would nevertheless be in their interest, in the sense of contributing to an important aspect of their well-being (e.g., their physical health). Importantly, this interpretation of interests allows for the possibility that a person can have a right to do or be provided with something even if they have no desire to do or be provided with that thing.

Since people have an interest in performing many kinds of actions and being provided with many kinds of goods and opportunities, we need a system by which we can distinguish between interests that do and do not give rise to rights-claims—otherwise, the interest theory is likely to lead to a troubling "proliferation of rights," where even the most trivial interests can be presented as the basis for valid rights-claims.[11] How, then, do we determine whether an interest is sufficiently important to justify holding others to be under a duty? There are generally two sets of considerations that must be taken into account. The first consideration is the centrality of the interest to a person's well-being. Other things being equal, the more central that a particular interest is to a person's well-being, the stronger the case will be for holding others to be under a duty with respect to that interest. Joel Feinberg offers a useful framework for categorizing our various interests by thinking of them as comprising an interconnected network that can be advanced or set back in different places.[12] We can imagine this network as being mapped onto a series of concentric circles, with our more important interests closer to the center and our less important interests closer to the periphery. At the center of our interest networks are what Feinberg calls our *welfare interests*, or interests whose satisfaction is a necessary precondition of pursuing valuable goals and projects. These include our interest in the continuation of our lives for the foreseeable future, in minimal physical health and rigor, in the integrity and normal functioning of one's body, in the absence of extreme pain and disfigurement, in basic subsistence and material security, and in freedom from unjustified interference and coercion, among others. Further from the center of our interest networks are what Feinberg calls our *ulterior interests* (which I will call our *project-based interests*) or the interests that we have in pursuing the goals and projects that give our lives meaning. These include our interests in making important artistic or scientific contributions, in advancing in our careers, in developing and maintaining meaningful relationships, in serving our communities, in engaging in leisure and sport, or in achieving spiritual enlightenment. Further still from the center of our interest networks are what we might call our *discretionary interests*, or the non-essential interests we have in the various good things in life. These may include our interests in eating delicious meals, in taking relaxing vacations, in accumulating wealth or property, or in enjoying the recognition and admiration of our peers.

Welfare interests are prime candidates for grounding rights-claims because the non-satisfaction of these interests will undermine or even fully prevent the satisfaction of interests elsewhere in our networks. For example, we cannot advance in our careers or maintain meaningful relationships unless we have reasonable assurance of our basic freedom and the continuation of our lives for the foreseeable future, and in this sense, rights to goods like personal security and subsistence are among the most basic of rights-claims. However, because the satisfaction of our welfare interests is only necessary, and not sufficient, for a flourishing life, there is also a strong case to be made for rights that protect our ability to pursue the various goals and projects that give our lives meaning, including rights to freedom of association, freedom of religion, freedom of occupational choice, and freedom of expression. Notice, however, that as we venture further from the center of our interest networks, the interests start to become more trivial and hence the cases for protecting them more tenuous, to the point where it can no longer be claimed that a particular interest is sufficiently important to justify holding others to be under a duty. Making these judgments is not without controversy—there is likely to be reasonable disagreement about the classification of different interests—though the concept of an interest network offers a useful framework in which to negotiate these disagreements and helps to explain why our interest in basic subsistence differs from our interest in eating gourmet meals with respect to its ability to ground a rights-claim.

The second consideration that must be taken into account when evaluating whether a particular interest is sufficiently important to ground a rights-claim is the nature and demandingness of the corresponding duties. Some interests are deeply important for a person's well-being, but protecting them as rights would require the imposition of duties on others that are unreasonably demanding. Take, for example, the interest we have in sexual intimacy. This interest is by no means trivial—it is arguably important enough to be considered a basic physical and emotional need—though few would endorse the idea that protecting this interest justifies holding others to be under a duty to provide their sexual services, as such a duty would require the sacrifice of interests on behalf of the duty-bearers that are too important to require them to give up, including their interests in bodily integrity or sexual autonomy.[13] Other things being equal, then, the more demanding the corresponding duty, the weaker the case will be for recognizing a particular interest as grounding a rights-claim. It is important to note, however, that what we consider to be excessively demanding will be closely tied to the importance of the interest in question, such that we are willing to accept the imposition of more demanding duties for the sake of protecting more fundamental interests.

Two additional points are worth noting about the demandingness of duties. First, because it is normally easier to passively abstain from interfering in another person's conduct than it is to actively provide them with assistance, positive rights will often face a higher justificatory threshold on the interest theory of rights. The point should not be overstated—there can be costs associated with abstention just as there can be considerable moral urgency associated with the provision of goods or opportunities—but in general terms, the duties associated with positive rights-claims will often be more demanding. Second, the demandingness associated with particular duties can also be context-dependent, such that the protection of a particular interest may justify the imposition of duties in some contexts but not others. For example, in normal circumstances, your interest in basic subsistence may justify the imposition of a duty on me to provide you with a portion of my food reserves, though this might not be true in the context of a famine, where the costs associated with discharging that duty—to myself or my family—may be considerably higher.

When both the capacity and sufficiency conditions are met, we can say that X has a right. What role should this play in our moral reasoning? According to Raz's own interpretation, rights that are justified on the basis of the interest theory carry "peremptory force," which is to say that they have the power to override potentially competing moral considerations. This interpretation aligns with other prominent interpretations of rights, in which they have been presented as "trumps"[14] or "side-constraints"[15] on the actions or decisions that are morally available to others, though it ought to be resisted for at least two reasons. First, as Jeremy Waldron has pointed out, rights that are justified on the basis of the interest theory are bound to conflict with each other, both in terms of *intra*-right conflicts (when the duties correlative to the right of one person conflict with the duties correlative to the same right of another person) and *inter*-right conflicts (when the duties correlative to one right conflict with the duties correlative to a different right).[16] Suppose, for example, that A and B both have an interest in health that is sufficiently important to justify a duty on behalf of the government to provide emergency medical care, though a scarcity of resources means that only one of them can be treated at a particular time. In this case, the duties correlative to A's right to health directly conflict with the duties correlative to B's right to health, such that neither carries peremptory force—on the contrary, it seems like we may have to draw upon other moral considerations in order to resolve the conflict, such as a principle of fairness (e.g., a coin flip) or of maximizing benefit. Or suppose that we recognize both a right to free speech and a right to personal security. We might then face a conflict in an instance where A's speech-act presents a serious risk of harm to B, e.g., by inciting an angry crowd to commit violence against him. In this case, making a determination about which right carries peremptory force seems to require what Waldron, following John Rawls, calls an order of lexical priority, or a ranking of different rights in order of their importance.[17] A tiered classification of interests might get us part of the way toward arriving at such a ranking (e.g., if we take the welfare-based interest in personal security to have lexical priority over the project-based interest in free speech), though there will certainly be cases in which the appropriate ranking of different rights is much less clear.

A second and more familiar reason to resist Raz's strong interpretation of rights is that discharging the duties correlative to a particular right can sometimes conflict with what one morally ought to do, all things considered. This is clearest in cases where discharging the duties correlative to a particular right conflicts with significant non-rights-based moral considerations, such as beneficence. Imagine, for example, that we can significantly improve the lives of five gravely ill children by reallocating a supply of medicine that is required to keep an elderly patient alive for a few more days. In this case, unless we think that considerations of rights are exhaustive of morality, it seems like we have a weighty moral reason to reallocate the medicine and therefore infringe on the elderly patient's right to life-saving medical treatment. Strict deontologists might resist this view on the grounds that rights must never be violated regardless of consequence, though this view becomes less plausible as the corresponding benefit increases (e.g., if we could significantly improve the lives of one million gravely ill children rather than five). In light of these considerations, it is perhaps more plausible to think of rights justified on the interest theory as strong *pro tanto* moral claims that are subject to being overridden in light of competing moral considerations, both rights-based and non-rights-based. This account is less neat and tidy than competing accounts of rights that stress the importance of "compossibility" (or mutual consistency) between different rights-claims,[18] though it is more reflective of the messy reality of moral and political life, where different values that we care about often seem to pull in different directions.

1.3 The Nature of the Interest in Procreation

Having outlined some of the basic features of the interest theory of rights, our main question can now be stated as follows: do individuals have an interest in procreation that is sufficiently important to justify holding others to be under a duty? Many commentators have been quick to respond with a resounding "yes." John Robertson, for example, holds that the right to procreate is justified because the ability to make one's own decisions about procreation is "central to personal identity, to dignity, and to the meaning of one's life,"[19] while John Harris has claimed that the right to procreate "expresses something so basic and deep-rooted in human psychology and social practice that it seems hardly worthy of special attention."[20] Robertson and Harris are not wrong in assuming that procreation involves fundamental human interests, though we need to know more about the nature of these interests before we can determine (a) whether they are sufficiently important to justify holding others to be under duties and (b) what kinds of duties they actually entail. In this section, I consider a number of possible interests involved in procreative decision-making and conclude that only one of them—the interest in parenting—is capable of grounding a right to procreate in the way that we normally understand that right. This will set the stage for my discussion in the remainder of the chapter, where I will outline and defend a justification of the right to procreate that is ultimately grounded in the interest in parenting.

Before proceeding, however, it is worth introducing one more Razian distinction. In presenting the interest theory of rights, Raz distinguishes between what he calls *core* and *derivative* rights, or rights that are grounded directly in interests, and rights that are merely the by-product of other rights.[21] For example, my right to freedom of speech may be grounded directly in the interest I have in expressing myself and voicing my moral and political opinions, though my right to peacefully protest on the grounds of a legislature is merely derivative of my right to freedom of speech. The distinction between core and derivative rights is relevant to the present discussion because it helps us to distinguish two different strategies that might be used to justify the right to procreate. One is to ground the right to procreate directly in an interest in procreation (which I will term *direct justifications*), while the other is to derive the right to procreate from a different right that is grounded in a different interest (which I will term *indirect justifications*). I will argue that the right to procreate is best understood as deriving from a more fundamental right to parent, though before doing so, it is first necessary to show why common alternative accounts are unsatisfactory. I will demonstrate that each of these accounts suffers from at least one of four key problems: (1) it is based on an interest that is not sufficiently important to justify holding others to be under duties; (2) it entails the imposition of duties that are unreasonably demanding; (3) it does not necessarily entail a set of duties that protects the ability of the right-holder to create and raise their own biologically related child, and hence does not accord with our pre-reflective intuitions about the value of the right to procreate; or (4) it cannot explain the special significance that we often associate with restrictions on the right to procreate.

1.3.1 *Direct Justifications*

Let us begin by considering direct justifications. If we pursue a direct justification of the right to procreate, then we are attempting to show that individuals have a significant interest in procreation *as such* that is sufficiently important to justify holding others to be under a duty. By a significant interest in procreation *as such*, I mean an interest in the mere creation of biological offspring conceived independently of any ancillary interest in rearing those offspring. One way to test whether there is a significant interest in procreation as such is to consider a scheme of duties that *only* protected one's ability to participate in the creation of biological offspring. Consider, then, the case of sperm donation. Men who donate sperm are permitted to participate in the creation of biological offspring, but they are not permitted, nor are they expected, to participate in the rearing of those offspring. If the right to procreate was grounded directly in an interest in the creation of biological offspring, then it could presumably support a moral right to donate sperm as a means of participating in the creation of biological offspring. However, such a right seems implausible for at least two reasons. The first reason is that the interest in question does not seem sufficiently important to justify the imposition of duties on others. Participating in the creation of biological offspring via sperm donation is not a necessary condition of pursuing other valuable projects or aims, and so does not constitute a basic welfare interest, nor does it exhibit the features that we commonly associate with important project-based interests, such as being central to personal identity, to dignity, or the meaning of one's life. After all, the "project" in question normally consists in ejaculating into a collection receptacle that is then used by others to create a child who the sperm donor may never meet. The empirical literature on sperm donation affirms the view that this interest is not typically regarded as being very important. For example, a systematic review of the experiences of sperm donors found that while a variety of motives are often cited for donating sperm, including altruism, the desire to pass on one's genes, and the desire to test one's fertility, most potential donors would not choose to donate their sperm if they were not also financially compensated.[22]

Of course, one might object to these claims on the grounds that participating in the mere creation of biological offspring via sperm donation *can* in fact realize important project-based interests, even if these interests are not always recognized or articulated by the donors themselves. For example, some have argued there is significant value in the creation of life insofar as it involves exercising one's agency and autonomy to produce something with inherent value,[23] while others have argued there is significant value in contributing to humanity's gene pool insofar as it reflects a belief that reproducing one's genes will benefit the world.[24] If pursuing either of these objectives was regarded as central to one's identity, dignity, or the meaning of one's life, then participating in the mere creation of biological offspring via sperm donation may in fact realize important project-based interests. Even if this were correct, however, notice that positing a moral right to do so would still be implausible for a second reason, which is that the duties it would impose on others are unreasonably demanding. In order for a sperm donor to realize their purported interest in creating life or contributing to humanity's gene pool, their sperm must be used to fertilize an ovum, which must then be gestated for nine months, after which the resultant child must be birthed and then raised for the better part of two decades into an independent adult. Whatever interests are at stake in the creation of biological offspring via sperm donation, they are not sufficiently important to justify the imposition of such a demanding set of positive duties on others. At best, they may be able to ground a negative right to non-interference or non-discrimination in the donation of sperm, though this would of course be insufficient for realizing the interest in question.

A similar line of reasoning can be extended to the case of gestational surrogacy, albeit with some important qualifications. Like sperm donors, gestational surrogates are also permitted to participate in the creation of biological offspring, but they are not permitted, nor are they expected, to participate in the rearing of those offspring. Unlike sperm donors, however, it is plausible to think that there is a weightier set of interests at stake in the ability to experience pregnancy and childbirth. Becoming pregnant, gestating a fetus, and giving birth to a child have been described as physically and epistemically transformative experiences that can hold significant value for those who undertake them.[25] For some women, the experience of pregnancy is regarded as "a foundational experience of being a *woman*," and the ability or inability to access that experience "could greatly impact [a woman's] conception of her life, its purpose and the distinct character or shape of her life experience."[26] Given these features, we might think that—unlike the interest in donating sperm—the interest in experiencing pregnancy and childbirth is at least the *type* of interest that can ground a rights-claim. Even if this is correct, however, the notion of a right to experience pregnancy or childbirth *simpliciter* is still implausible in light of the duties it would impose on others. In order for a gestational surrogate to satisfy their interest in gestating and giving birth to a child, they must be provided with others' reproductive gametes with which to create an embryo to gestate (at least sperm, though potentially an ovum as well), and more importantly, someone else must assume the responsibility of raising the resultant child into an independent adult. Even if we accept that there are non-trivial interests involved in the experience of pregnancy or childbirth, they are not sufficiently important to justify the imposition of such a demanding set of positive duties on others. As in the case of sperm donation, they could at best support a negative right to non-interference or non-discrimination in the opportunity to serve as a gestational surrogate, though this would again be insufficient for realizing the interest in question.

A final reason why direct justifications of the right to procreate are implausible is because they do not necessarily generate a set of duties that protect one's ability to create and raise their own biological child, and hence do not accord with our pre-reflective intuitions about the value of the right to procreate. Imagine, for example, a society modeled after Plato's *Republic*, in which children are raised collectively in state-run orphanages.[27] Apart from designated caregivers, citizens are not permitted to participate in the rearing of children, but they can participate in the creation of children by either donating their reproductive gametes to a central bank or, in the case of biologically female citizens, serving as gestational surrogates. Unlike our current society, this society can fully accommodate a right to procreate that is grounded directly in an interest in the creation of biological offspring, as its institutional structure allows for the realization of that interest without imposing duties on others that are excessively demanding. However, few (if any) prospective procreators would prefer such a society to our own, suggesting that an interest in the mere creation of biological offspring is not what ultimately grounds the right to procreate.

1.3.2 *Indirect Justifications*

The failures associated with direct justifications of the right to procreate should lead us to consider indirect justifications, which seek to derive a right to procreate from other rights grounded in other types of interests. Given its close connection to the body and bodily processes, it is tempting to argue that the right to procreate can be derived from a more general right to bodily integrity, where the right to bodily integrity is understood as the right to make one's own decisions about the uses of one's body. Christine Overall, for example, argues that the right to procreate is ultimately grounded in individuals' (particularly women's) bodily freedom and autonomy, suggesting that "without moral and legal recognition of their bodily freedom and autonomy, women are little more than procreative slaves."[28] Laura Shanner has advanced a similar argument, suggesting that the various activities that are normally thought to be protected by a right to procreate are in fact protected by rights to bodily integrity. As she claims,

> Most reproductive rights concerning abortion, sterilization, and contraception are negative rights justified by appeal to more general negative rights to freedom from physical assault or coercion, especially by the state ... We need not appeal to any specific *reproductive* rights to challenge such interventions; we need merely to point to basic rights to be free from bodily invasion by governments or doctors.[29]

Unlike an interest in the mere creation of biological offspring, an interest in making one's own decisions about the uses of one's body is sufficiently important to justify the imposition of duties on others not to interfere in those decisions, at least insofar as they do not infringe on the legitimate moral claims of others. But can a right to bodily integrity support a derivative right to procreate? The right to bodily integrity plausibly supports a derivative right *not* to procreate, as there is generally no way of forcing a person to procreate against their will that does not involve invasive interferences in their body. It may also support certain forms of non-interference in decisions *to* procreate, for, as Shanner notes, historically significant forms of interference in procreative decision-making have typically involved serious violations of bodily integrity, including forced abortion, forced sterilization, and forced contraceptive use. However, a key limitation of a bodily integrity-based justification of the right to procreate is that it cannot explain why non-bodily forms of interference in procreative decision-making are objectionable. This becomes especially problematic when we consider that, for many persons, including those who are single, infertile, or non-heterosexual, respect for bodily integrity is insufficient for realizing their interest in procreation. Imagine, for example, a government policy that prohibited access to donated gametes or fertility treatment for persons who are single, infertile, or non-heterosexual. If the right to procreate was understood solely as a by-product of the negative right to bodily integrity, then there would be no objection to this policy based on the right to procreate. This, however, seems implausible. As Overall herself acknowledges, given the significant interest that most people have in creating and raising their own biological children,

> access to a variety of treatments intended to support pregnancy and birth and to enhance, repair, or restore fertility ... should not be denied based on social identity characteristics such as marital status or sexual orientation, nor should it be denied on the basis of age, health status, or impairment.[30]

This is certainly correct; though if we think the interest in procreation weighs against non-bodily forms of interference in procreative decision-making, this suggests that it is not reducible to our interest in bodily integrity, despite the obvious importance of that interest in the domain of procreative decision-making more generally.

A second indirect justification derives a right to procreate from a more general right to personal autonomy or self-determination. Dan Brock, for example, argues that "the primary moral basis of reproductive freedom is individual or personal autonomy," where an individual's interest in autonomy is understood as "their interest in making significant decisions for themselves according to their own values or conception of a good life."[31] It is not difficult to see how procreation is the type of decision that might be protected by a general right to personal autonomy. As Brock claims,

> whether one is free to procreate at all affects the fundamental issue in many people's lives of whether having and raising children will be part of their lives. Since for most parents this is a central, if not the central, project in their lives, the decision about whether to have children at all is of fundamental importance to them.[32]

One of the advantages of an autonomy-based justification of the right to procreate is that it is capable of explaining why both bodily and non-bodily forms of interference in procreative decision-making are objectionable. Just as one's autonomy is frustrated when one is prevented from having biological children via coercive sterilization, so too is one's autonomy frustrated when one is prevented from having biological children due to arbitrary restrictions on one's access to gametes or fertility treatment.[33] However, one significant drawback of deriving a right to procreate from a general right to autonomy is an inability to account for the special significance of restrictions on one's autonomy that impact procreative decision-making. Governments regularly restrict our autonomy in a variety of ways, though our reaction to these restrictions varies according to the nature of the actions or decisions that are being restricted. Compare, for example, the following two policies: (1) a zoning restriction that prevents you from building your dream home on a plot of land that you own, and (2) a restriction on your ability to access the gametes or fertility treatment you require to procreate. Both of these policies restrict your ability to make a significant decision for yourself based on your own values or conception of the good life, though most people would regard the second policy as imposing a much more serious set of restrictions on your autonomy than the first. Notice, however, that we cannot account for this difference without saying something more specific about the nature and significance of the activity that is being restricted. Thus, like the bodily integrity-based account, the autonomy-based account of the right to procreate is incomplete: while procreation may indeed be the type of activity that is protected by a general right to personal autonomy, appealing to autonomy alone cannot explain the special significance of restrictions on autonomy in the context of procreative decision-making. In order to fill this gap, we need a more detailed explanation of *why* procreative decision-making is such an important exercise of personal autonomy, such that it constitutes "a central, if not the central, project"[34] in the lives of many procreators. The most obvious explanation, hinted at by Brock himself, is supplied by a third indirect account of the right to procreate, which derives a moral right to procreate from a more fundamental right to parent.

1.4 The Parental Interest Justification of the Right to Procreate

I believe this account provides the most promising justification for the right to procreate, so it is worth exploring in some detail. To do so, I will first outline the case in favor of recognizing a moral right to parent and will then explain how, and under what conditions, a moral right to procreate may be derived from it.

Harry Brighouse and Adam Swift have advanced the most fully developed case for recognizing a moral right to parent based on the distinctive contribution that parenting makes to the well-being of the adults who undertake it.[35] Their account begins with the observation that intimate relationships are necessary for our lives to go well. However, not all relationships are the same, nor are they necessarily substitutable, and we tend to require a variety of relationships in order to flourish. For example, we need close friends with whom we can confide and go to for emotional support; casual acquaintances with whom we can bond over shared interests; and romantic partners with whom we can reveal ourselves fully and share sexual intimacy. In this sense, a successful life will normally include a variety of successful relationships that can serve our various relational needs.

Another type of relationship that many adults require to flourish is the intimate relationship that develops between parents and their children. According to Brighouse and Swift, this is not just another intimate relationship, one that is valuable to the adults who undertake it but ultimately substitutable by an additional relationship with a consenting adult. Rather, it is "a relationship that involves the adult in a unique combination of joys and challenges; experiencing and meeting these makes a distinct set of demands on him and produces a distinctive contribution to his well-being."[36] Part of the uniqueness of the parent-child relationship consists in its strongly *fiduciary* character. In becoming a parent, an adult agrees to secure the conditions for the child's immediate well-being and to oversee their long-term physical, cognitive, emotional, and moral development.

Performing this role can be a source of deep satisfaction and offers numerous opportunities for personal development. However, as one can serve in a fiduciary role with respect to another adult, such as an elderly family member or an adult with a cognitive disability, the uniqueness of the parent-child relationship cannot wholly be explained by its fiduciary character. On the contrary, Brighouse and Swift identify four additional features of this relationship that combine to make it both welfare-generating and non-substitutable, and hence distinctively valuable for adults. First, unlike relationships between adults, the relationship between parents and children is deeply structurally unequal, with children being especially vulnerable to and dependent on the decisions of their parents. Second, the parent-child relationship is always characterized by a form of legitimate paternalism, wherein parents are required to govern children in ways that will promote their immediate and long-term interests. Third, parent-child relationships are also unique in that children lack a determinate conception of their own good, such that parents cannot simply act on the basis of the child's own beliefs about what is good for them (as in other fiduciary relationships), but must also shape their beliefs about what is valuable in life. Finally, and perhaps most importantly, parent-child relationships are also distinct in the unique brand of intimacy they make possible, wherein the "love that one receives from one's children ... especially in the early years, is spontaneous and unconditional and, in particular, outside the rational control of the child."[37] According to Brighouse and Swift, adults have a strong interest in being able to occupy a relationship characterized by these features. As they claim,

> The [parental] role enables [parents] to exercise and develop capacities the development and exercise of which are, for many (though not, certainly, for all), crucial to their living fully flourishing lives. Through exercising these capacities in the specific context of the intimately loving parent-child relationship, a parent comes to learn more about herself, she comes to develop as a person, and she derives satisfaction that otherwise would be unavailable.[38]

For Brighouse and Swift, the strong interest that adults have in accessing the distinctive goods associated with parent-child relationships is sufficiently important to ground a moral right to parent. Importantly, the acquisition and maintenance of this right is conditional on the adult discharging their fiduciary duties to an adequate standard—an adult who cannot or will not do so does not have a right to become or remain a parent.[39]

Of course, even if we agree that an interest in parenting is sufficiently important to ground a moral right to parent, we still might wonder: what, exactly, is the right to parent a right *to*? That is, what is the content of the duties such a right imposes on others? Given that one cannot realize their interest in parenting without having access to a child to parent, it might be tempting to think that a moral right to parent must entail a positive right to be provided with a child to parent. On this interpretation, the moral right to parent is really a right *to* a parent-child relationship. However, this interpretation of the right to parent is implausible for numerous reasons, which can be illustrated through the following case:

> *Single Man*: Jack is a single man living in a small community of families. Seeing the other adults in his community flourish as a result of their participation in parent-child relationships, Jack desperately wants to become a parent, yet he is unable to find a romantic partner with whom to create a child to parent and there are no children in his community who are currently parentless and in need of adoption.

In order for Jack to realize his interest in parenting, one of two things must happen: either (1) a child must be created for him to parent, or (2) an existing child must be reallocated to him to parent. However, neither can plausibly be construed as duties correlative to Jack's right to parent, even if we grant that his interest in parenting is quite weighty. The duty outlined in (1) is too demanding to be morally required, for while Jack's interest in parenting is certainly important, it is not sufficiently important to justify the imposition of a duty on one of the women in his community to conceive, gestate, give birth to, and ultimately relinquish a child for him to parent. The duty outlined in (2) is also problematic, as the reallocation of an existing child for Jack to parent directly conflicts with the interest of that child's parents in maintaining a relationship with the child—an interest that is protected by *their* right to parent—as well as the child's own interest in continuity of care.[40]

The moral right to parent is thus better understood as a right to non-interference and/or positive assistance in the initiation and maintenance of parent-child relationships. On this view, Jack is not morally entitled to a parent-child relationship, but he is entitled to non-interference in the pursuit of his interest in parenting and perhaps also to limited forms of positive assistance that do not exceed a demandingness threshold. Brighouse and Swift are particularly concerned with the negative dimension of the right to parent and how it might be used to explain the wrongness of laws or social policies that impose obstacles to one's ability to realize their interest in parenting. For example, they famously imagine a hypothetical policy of "baby redistribution," in which an efficient and well-run government agency reallocates babies at birth to persons who are judged most likely to raise them best, thus leaving some adequate would-be parents childless. Even if such a system genuinely worked to the advantage of children, Brighouse and Swift think that we would still have reason to object to it insofar as it interferes with the ability of some persons to access the goods associated with parent-child relationships.[41] We can also imagine more realistic scenarios in which the right to parent has a similar application. For example, it might be invoked to frame a moral objection to a discriminatory adoption system that withholds adoptable children from single individuals or same-sex couples who would otherwise be competent and loving parents. Such a system arguably wrongs those individuals by interfering in their ability to initiate parent-child relationships (in addition to wronging the children who would otherwise benefit from those relationships).

If the right to parent is best understood as a right to non-interference or even positive assistance in the initiation and maintenance of parent-child relationships, does it also support a moral right to create a child as a means of initiating a parent-child relationship? Brighouse and Swift do not explicitly consider whether the moral right to parent also supports a moral right to procreate, though they clearly see a close connection between these two ideas. For example, in discussing the possible duties correlative to the right to parent, they suggest that such a right entails "a right not to be prevented from parenting the children one has procreatively produced" and may also entail positive claims to the provision of fertility treatment as a means of creating children with whom one can initiate parent-child relationships.[42] In a brief but more direct comment on the relationship between procreation and parenting, they also note that "the alleged 'right to procreate' derives much of its appeal from an implicit claim about the value of getting to parent the child one has procreatively produced,"[43] suggesting a justification of the right to procreate that is grounded in a more fundamental interest in parenting. However, while Brighouse and Swift remain noncommittal about the implications of their view for the morality of procreation, there is good reason to believe that a moral right to procreate follows directly from their account of the moral right to parent. This claim is supported by the following line of reasoning, which I will henceforth refer to as the *parental interest justification of the right to procreate*:

1 adults have an interest in parenting that is sufficiently important to ground a right to non-interference or even positive assistance in the initiation and maintenance of parent-child relationships;
2 in order to initiate a parent-child relationship, one must have access to a child in need of parental care;
3 the primary and in some cases only way to permissibly gain access to a child in need of parental care is to create a child; therefore,
4 an interest in parenting is sufficiently important to ground a right to non-interference or even positive assistance in the creation of children with whom one can initiate parent-child relationships, at least insofar as doing so is necessary to realize the goods associated with parent-child relationships.

This argument follows a similar justificatory strategy as other derivative rights, namely, individuals have a right to *x*; *y* is a necessary condition of achieving *x*; therefore, individuals also have a derivative right to *y* as a means of achieving *x*. Of course, as the qualification in (4) indicates, it is not *always* necessary to create a child in order to enjoy the goods associated with parent-child relationships—one can also enjoy those goods by adopting a child in need of parental care—and so an interest in parenting may not *necessarily* justify a right to procreate in all instances. However, while one can access the goods associated with parent-child relationships by adopting a child in need of parental care, it is important to note that this will not always be possible due either to the unavailability of adoptable children or the considerable costs that are associated with adopting a child, at least in current, non-ideal conditions.[44] Moreover, as Sarah Hannan and R.J. Leland have pointed out, every child who is parented must be procreated, and "even if every person who procreated but preferred not to parent the resulting children gave those children up to adults with a strong interest in parenting, the demand would likely outstrip the supply."[45] Thus, while an interest in parenting will justify a right to procreate only insofar as this is necessary to access the goods associated with parent-child relationships, this condition will often be met, even if it will not always be met.

At first glance, the parental interest justification of the right to procreate appears to avoid the problems associated with the alternative justifications considered so far. First, unlike the interest in the mere creation of biological offspring, the interest in parenting a child is sufficiently important to justify the imposition of duties on others given the distinctive contribution that parenting a child makes to adults' well-being. Second, when understood properly as a right to non-interference and/or limited forms of positive assistance in the creation of children with whom one can initiate parent-child relationships, the right to procreate does not entail the imposition of duties on others that are excessively demanding, such as duties to gestate and give birth to children or duties to raise the children that others have created into maturity. Third, a right to procreate grounded in an interest in parenting would not be consistent with the reallocation of children to state-run orphanages and thus better aligns with our pre-reflective intuitions about the value of the right to procreate. Finally, unlike justifications of the right to procreate derived from rights to bodily integrity or personal autonomy, the parental interest justification of the right to procreate is capable of supporting the procreative rights-claims of persons who cannot procreate without assistance, and it can account for the special significance that is often associated with restrictions on one's autonomy in the domain of procreative decision-making.

Despite these advantages, however, the parental interest justification of the right to procreate is also vulnerable to at least three important objections, each of which stems from one of its implicit assumptions about the moral relationship between procreation and parenting. The first challenges the assumption that procreation is a permissible or unproblematic means of initiating a parent-child relationship; the second challenges the assumption that procreators are entitled to enter into parental relationships with the children they bring into existence; and the third challenges the assumption that the value of procreation is reducible to its function as a means of accessing the goods of parenting. Some of these challenges will be addressed more fully in later chapters, when I discuss, for example, the conditions under which it is morally permissible to bring a child into existence (Chapter 2), the relationship between procreation and the acquisition of parental rights (Chapter 3), and the moral significance of biological connectedness in debates about parental obligation (Chapter 4) or adoption (Chapter 6). Nevertheless, I would like to briefly outline them here and at least indicate the type of response that is required in order for the parental interest justification of the right to procreate to succeed.

1.4.1 *Is Procreation a Morally Permissible Means of Initiating a Parent-Child Relationship?*

The parental interest justification of the right to procreate assumes that creating a child is a morally permissible means of initiating a parent-child relationship. However, this assumption might be challenged: one could grant that it is permissible to initiate a parent-child relationship with an existing child in need of parental care in light of the benefits it would confer on that child, while maintaining that it is impermissible, or at least morally problematic, to create a child for that purpose. Hannan and Leland have proposed a version of this argument. It is based on the idea that the distinctive goods associated with parent-child relationships are based on children's subjection to experiences and states of affairs that are bad for them, including an impaired capacity for practical reasoning, the need for extensive parental control, profound and asymmetric vulnerability, and a lack of an established practical identity.[46] According to Hannan and Leland, these features of childhood are generally bad for children, but they are also inextricably linked with the goods of parenthood, as many of the goods associated with raising children involve helping them to cope with and eventually overcome these deficits. For example, one of the distinctive goods that parents enjoy as a result of raising children is the sense of responsibility and purpose that attends advancing their interests and protecting them from harm, but this is only possible because children are in a state of profound vulnerability with respect to others' actions and decisions. This state of profound vulnerability is something that is generally bad for children, as it leaves them in a position of complete dependence where they are persistently at risk of harm.

If we accept that many features of childhood are in fact bad for children, then the parental interest justification of the right to procreate begins to look problematic, as we are not typically permitted to put others into bad states just to satisfy our own interest in helping them to overcome them. Hannan and Leland support this claim with a series of analogies:

> Suppose that a person would genuinely flourish by nursing a close friend back to health as a result of the intimacy that such care would foster. This in no way speaks in favour of creating a situation in which her friend is ill and in need of care. Similarly, if a person would benefit from helping a romantic partner through a difficult time, this doesn't provide her with good reason to engineer difficulties for her partner. While each caregiver's significant interest in special relationships would ordinarily give strong reason to create a situation in which they enjoy those relationships, this reason is undercut by the fact that the interest essentially requires putting the recipients of care in seriously bad states.[47]

According to Hannan and Leland, parallel reasoning applies in the case of procreation and serves to undermine the parental interest justification of the right to procreate. While an interest in accessing the goods associated with parent-child relationships initially appears to speak strongly in favor of a right to create children with whom one can enjoy that type of relationship, it ultimately fails to do so because "the interest's satisfaction requires directly that children be non-consensually subjected to some very bad states."[48]

How should a proponent of the parental interest justification of the right to procreate respond to this objection? The first and perhaps simplest line of response is to deny the premise that childhood is bad for children, or at least to point out that its bad-making features are offset by its good-making features. For example, while children may experience a range of deficits as a result of being children, many commentators have observed that they also enjoy privileged and in certain cases exclusive access to a variety of goods, including unstructured time, imaginative play, intense experimentation, and sexual innocence.[49] These goods speak in favor of childhood as a uniquely valuable life stage, rather than a regrettable period of deficiency. Indeed, if childhood was regarded *purely* as a period of deficiency, then it would seemingly be rational to skip over it if we had the opportunity to do so, e.g., if, as Samantha Brennan imagines, we had access to a pill that would cause children to grow up instantaneously.[50] However, many would regard the bypassing of childhood to be a significant loss, which casts doubt on the view that childhood is bad for children; and if childhood is not bad for children—or at least not *all* bad for children—then creating a child for the purpose of parenting is not akin to putting someone in a bad state just to satisfy one's own interest in helping them overcome it.

Of course, simply acknowledging the positive aspects of childhood in no way erases its negative aspects, nor does it necessarily impact the claim that many of the goods associated with parenting are dependent on children experiencing those negative aspects. Moreover, it is precisely this latter claim that Hannan and Leland rely on to advance their view. Thus, a second line of response is to concede the badness of childhood (or at least its bad aspects), but to deny that this renders the creation of children wrongful in light of the benefits it also makes possible. Appealing to the benefits of existence—and the necessity of passing through childhood in order to access those benefits—marks a morally relevant disanalogy between the case of creating a child to enjoy the goods of parenting and the case of making a friend sick to enjoy the goods of caring for her. When we make a friend sick to enjoy the goods of caring for her, we are not delivering an overall benefit to her, never mind one that requires her subjection to the negative aspects of being sick—rather, we are simply making her worse off with respect to a baseline of health and then returning her to that very same baseline. However, the same is not necessarily true when we bring a child into existence to satisfy our interest in parenting. Even though we must always bring a child into a state that is in certain respects bad for them, experiencing this state is necessary to enjoy whatever benefits are associated with existing once this state is overcome (if we think there are such benefits), including developing meaningful relationships, pursuing valuable goals and projects, experiencing arts and culture, or witnessing the beauty of the natural world.[51] Thus, even if we accept that childhood is a bad state, or is characterized by certain bad-making features, there appears to be a justification for children's subjection to the bads of childhood that does not apply to a friend's subjection to the bads of being sick, suggesting that our objection to the latter type of case does not necessarily implicate the former. Of course, there may be cases in which a child's existence is *not* expected to be beneficial in these ways, in which case an appeal to these benefits could not justify their subjection to the bads of childhood. I will address these types of cases in Chapter 2, when I consider how the interests of prospective children constrain the conditions under which it is morally permissible to procreate. However, I will assume for now that the presence of the benefits associated with existence suffices to render children's subjection to the bads of childhood non-problematic, and in a way that renders the dependence of the interest in parenting on the bad-making features of childhood similarly non-problematic.

1.4.2 Is There a Right to Parent One's Biological Child?

Let us assume, then, that we do not necessarily act wrongly when we bring a child into existence as a means of initiating a parent-child relationship.

A second objection one might raise against the parental interest justification of the right to procreate is that it assumes, without argument, a right to enter into parental relationships with the children one has brought into existence. The legitimacy of such a right is normally taken for granted in law and common-sense morality, though there are valid reasons for questioning it, particularly if we buy into Brighouse and Swift's account of the importance of the interest in parenting. To understand the nature of these reasons, let us return for a moment to the case of *Single Man*. Imagine that one of the couples in Jack's community—the Smiths—are already parenting several children and are soon expecting another. While they are capable of fulfilling their parental duties to this child to an adequate standard, they are utterly exhausted with their existing parental responsibilities and are likely to do a worse job than other prospective parents who have less on their plate. Imagine further that Jack has all of the character traits of a loving and devoted parent—in fact, he is by all accounts the best available parent in the child's community. Do the Smiths nevertheless have a right to parent their biological child?

Existing approaches toward the just distribution of parental rights suggest otherwise. For example, if we take a *child-centered* approach and assume the right to parent should be allocated solely on the basis of what is best for the child, then our answer to this question is "no": Jack is better positioned to advance the child's interests than the child's procreators, and so, other things being equal, the right to parent should be allocated to him. If, on the other hand, we take a *dual-interest* approach and assume that the allocation of the right to parent must also take into account the interests of adults in parenting (as Brighouse and Swift do), then our answer is again quite plausibly "no": Jack's interest in becoming a parent is weightier from a moral point of view than the Smiths' interest in parenting an additional child, and so the right to parent should again be allocated to Jack. Indeed, by allocating this right to Jack, we seem to be killing two birds with one stone, promoting a more equitable distribution of parenting opportunities and optimizing the quality of parental care that is available to the child.

This type of case presents a challenge for the parental interest justification of the right to procreate, as it calls into question whether there is in fact a right to enter into parental relationships with the children one has brought into existence. If there is no such right, then procreation is not necessarily a means of permissibly gaining access to a child to parent, and hence an interest in parenting will not necessarily justify a right to create a child. How should a proponent of the parental interest justification respond to this challenge? One option, of course, is simply to concede the challenge: perhaps the right to parent one's biological child is actually baseless, and justice in fact requires assigning the right to parent on the basis of other types of considerations. Depending on what these considerations are, we may sometimes be required to reallocate children away from their biological progenitors to more deserving adoptive parents. This option would of course undermine the parental interest justification of the right to procreate by limiting the conditions under which premise (3) is true.

The second option is to show that a child's procreators normally satisfy whatever conditions we take to be relevant for the allocation of the right to parent. There are at least two variations on this option, depending on the approach we take toward the just distribution of parental rights. If we take a child-centered approach, then we can attempt to show that a child's procreators are normally best positioned to serve as their custodial parents. Several philosophers have defended versions of this view. For example, Anca Gheaus has argued that a child's gestational mother will normally satisfy the condition of being the best available parent given the intimate bonding that occurs between babies and their birth mothers during the process of gestation,[52] while David Velleman has argued that children have an interest in being parented by their procreative parents that is derived from their interest in having a connection to their ancestral history.[53] Alternatively, if we take a dual-interest approach, then we can attempt to show that a child's procreators normally have a weightier interest in parenting the children they have brought into existence than other prospective parents, such that they are uniquely entitled to enter into parental relationships with their biological children. I will defend a version of this dual-interest strategy in Chapter 3, where I argue that procreators have a presumptive right to parent the children they bring into existence insofar as the assumption of a parental role with respect to those children can be understood as the continuation of a justice-respecting parental project. If successful, this account forms part of a mutually supportive package of rights protecting one's ability to create and raise a biologically related child: the interest in accessing the goods associated with parent-child relationships entitles one to non-interference or even limited forms of positive assistance in the creation of a child to parent, while the just initiation of that project entitles one to non-interference in its continuation through the assumption of a parental role with respect to that child.

1.4.3 *Is Procreative Parenting Distinctively Valuable?*

The third objection to the parental interest justification does not challenge whether an interest in parenting can justify a right to procreate, but rather challenges whether the value of procreation is reducible to its function as a means of initiating a parent-child relationship. Proponents of this type of objection insist that there is distinctive value associated with parenting a child one has brought into existence, such that one's ability to pursue parenthood via procreation deserves special protection. Luara Ferracioli has advanced a version of this argument focusing on the unique quality of parental love that is made possible as a result of procreation.[54] According to Ferracioli, procreation has distinctive value because it provides reasons for the development of a particularly deep and robust brand of parental love, one that does not depend on the intrinsic qualities of the child or any relational history between the child and the parent. For example, procreators tend to develop deep and robust love for the children they bring into existence even when those children are difficult to raise and even when they lack a history of interaction, e.g., in the case of estranged biological fathers. Ferracioli argues that this type of love is the closest we can get to unconditional love and that an interest in experiencing it can ground a legitimate preference for procreative parenting. She summarizes her view in the following way:

> Procreative-parenting justifies love irrespective of facts about the child or facts about the relationship. The love of procreative-parenting is therefore the closest we can get to unconditional love: the mere fact that a parent has intentionally created her child gives the parent a weighty pro tanto reason for loving the child deeply and robustly. This, in turn, gives parents reasons to take on costs and make sacrifices for the good of their child across a lifetime. It is therefore not surprising that so many persons are subjectively attracted to procreative-parenting, and see it as an important ingredient of their conception of what it means for their lives to go well. They want to experience precisely this form of love, and in so doing, to care deeply and robustly about someone else's good irrespective of her intrinsic qualities and irrespective of facts about their mutual interactions over the course of an intimate relationship.[55]

If sound, Ferracioli's argument suggests two shortcomings of the parental interest justification of the right to procreate. First, it suggests that the interest in premise (1) is under-described to the extent that it refers to the interest in parenting in general and not to the specific interest in parenting a child one has intentionally brought into existence; and second, it suggests that the qualification in premise (4) is mistaken to the extent that we can only realize this interest by creating a child.

In order for this argument to succeed, however, it must be the case that the quality of parental love Ferracioli describes is exclusive to procreative parenting, such that it can *only* be enjoyed as a result of creating a child. Is this the case? Skeptics will likely point out that adoptive parents also develop deep and robust love for the children they adopt into their families. For example, just like procreative parents, adoptive parents will routinely sacrifice their own interests in favor of the interests of their children, and their love for their children will persist through significant changes in their values, ambitions, and life circumstances. In this sense, at least two of the valuable features that Ferracioli associates with parental love—namely, its *depth* and *robustness*—appear to be accessible to adoptive parents in equal measure.

Ferracioli does not deny that adoptive parents have reason to love their children deeply and robustly once they enter into parental relationships with them, but she insists that "such love will be justified by an appeal to facts about the child and/or facts about the relationship, and not by an appeal to the fact of adoption."[56] This is importantly distinct from procreation, which in her view justifies parental love independently of these factors. To illustrate this difference, she presents the following case:

> Ada has been on the waiting list for adoption for 9 months and has finally been assigned a 12-month-old child, Chelsea, whom she can take home after signing the adoption papers. However, after signing the final legal document, an orphanage worker rushes in and lets Ada know that a different child of a similar age, Cheryl, would be a much better fit for Ada (Ada's native language is French and Cheryl was cared by a French-speaking mother before becoming an orphan). She therefore suggests to cancel Chelsea's adoption so Ada can take Cheryl home instead.[57]

According to Ferracioli, it would be strange if Ada refused this offer on the grounds that, by signing the adoption papers, she had committed specifically to love *Chelsea* deeply and robustly. The strangeness of this response is explained by the fact that, in the case of adoptive parenting, love is not justified by the mere fact of adoption, but rather by facts about the child, facts about the relationship, or some combination of the two. By contrast, we would not find it strange were a procreator to choose to parent their biological child even if they had the opportunity to parent a different child with whom they would have a better relationship (e.g., by swapping children with another parent). This is explained by the fact that, in the case of procreative parenting, love is justified by the mere fact of procreation and remains so irrespective of the intrinsic qualities of the child or of their relationship's history.

For the sake of argument, let us grant Ferracioli's claim that, unlike adoptive parents, procreative parents have reasons for love that are untethered in these ways, such that there is a relevant difference between parenting via procreation and parenting via adoption. We still might wonder why this difference is valuable, such that it justifies affording a special status to procreative parenting. Imagine, for example, that Ada has decided to become a parent and is contemplating whether to pursue parenthood via procreation or adoption. Whatever option she takes, she knows that she will come to love her child deeply and robustly and will thereby experience the valuable features of parental love. What reason could we offer Ada for choosing to procreate? It would seem odd to point out that, in the case of procreation, she will be justified in loving her child independently of the child's qualities or their history of interactions, whereas in the case of adoption, her reasons for loving her child will be dependent on these features. After all, once she assumes a parental role in either case, she will develop relationship-based reasons to love her child deeply and robustly, and these reasons will become primary in justifying her love going forward—the fact that she would have independent reasons to love her child in the case of procreation does not seem particularly significant. Of course, it would be a different story if the love that procreators developed for their progeny was deeper or more robust than the love that adoptive parents developed for their adopted children, but there is no reason to believe this is the case.[58]

I assume in what follows that the valuable features of parent-child relationships are available to procreative and adoptive parents in equal measure, such that there is no distinctive value in procreative parenting that is being overlooked by the parental interest justification of the right to procreate. This is not to deny, however, that there may be ancillary interests involved in creating a child to parent. For example, in choosing to adopt rather than procreate, Ada will be forgoing the experience of creating life, of passing on her genes, of experiencing gestation and childbirth, and of parenting a child who physically resembles her. An interest in having these experiences can explain why someone might prefer to pursue parenthood via procreation rather than adoption, though it does not necessarily justify the accommodation of that preference, particularly in cases where there are moral considerations that weigh against it. As I will argue further in Chapter 6, the preference for procreative parenting in these types of cases might reasonably be regarded as something akin to an "expensive taste," or a discretionary interest whose protection does not necessarily justify the imposition of duties on others.

1.5 The Content and Limits of the Right to Procreate

So far I have argued that a moral right to procreate can be derived from a more fundamental right to parent, which is itself grounded in the strong interest that adults have in developing intimate fiduciary relationships with children. This in itself is not a particularly novel or surprising claim—after all, when undertaken intentionally, procreation is normally pursued with the explicit aim of creating a child to parent. However, adopting a parental interest justification of the right to procreate has some surprising implications for how we understand its content and limits. I would like to conclude my discussion by commenting briefly on the content of the right to procreate and the unique limitations that emerge when we embrace a parental interest justification.

Consider first the issue of content. The parental interest justification holds that an interest in parenting is sufficiently important to ground a right to non-interference or even positive assistance in the creation of children with whom one can initiate parent-child relationships, at least insofar as doing so is necessary to access the goods associated with parent-child relationships. Thus, a parental interest-based right to procreate can be understood to impose negative duties on others not to interfere in the creation of children with whom one can initiate a parent-child relationship, as well as positive duties of assistance that do not exceed a demandingness threshold.

This latter implication is controversial, as philosophers disagree about the positive dimensions of the right to procreate. For instance, while Overall suggests the right to procreate may entail a positive claim "to have a child by any means one may choose and to be provided with all possible assistance in reproduction,"[59] Robertson insists that the right to procreate is primarily "a negative right against private or public interference, not a positive right to the services or resources needed to reproduce."[60] A more defensible position can be found somewhere between these two extremes. One reason to resist a strictly negative interpretation of the right to procreate is that freedom from interference is normally insufficient for realizing one's interest in creating a child to parent. This is most obviously true in the case of the infertile, who may require access to fertility treatments like *in-vitro fertilization* (IVF) in order to bring a child into existence, but it is also true in the case of the fertile as well, who normally require access to basic obstetric healthcare in order to ensure a safe pregnancy and birth. Thus, if the right to procreate is meant to protect one's ability to achieve rather than merely attempt the creation of a child to parent, we should think of it as a right entailing both negative and positive duties.

At the same time, however, there are important limits on the nature and extent of the positive duties the right to procreate can plausibly entail. First, if we recognize individuals as having rights to bodily integrity, then the right to procreate cannot impose duties on others to provide their gametes or gestational services. A man does not have a legitimate claim to the ovum or gestational services of a non-consenting woman any more than a woman has a legitimate claim to the sperm of a non-consenting man, suggesting that positive claims to assistance will most plausibly be limited to "a variety of treatments intended to support pregnancy and birth and to enhance, repair, or restore fertility."[61] Second, even if positive assistance is limited to the provision of services like fertility treatment or obstetric healthcare, the extent of assistance the right to procreate is able to support will also be limited by a society's level of affluence as well as more general considerations of distributive justice. For example, whether or not the right to procreate supports a claim to IVF treatment will depend on whether a society is capable of offering that form of assistance without negatively impacting the just claims of other people, including their claims to limited health resources.

A parental interest-based right to procreate will therefore impose both negative duties of non-interference and limited duties of positive assistance in the creation of children with whom one can initiate parent-child relationships. What, then, are the limitations of this right? In addressing this question, it is helpful to distinguish between *internal* and *external* limitations to a right.[62] Internal limitations derive from the interest that grounds the right and specify the conditions under which that right may be appropriately exercised or claimed, whereas external limitations derive from competing moral considerations that limit the circumstances in which a right may be appropriately exercised or claimed. Consider, for example, the right to freedom of expression. This right is intended to protect a particular interest—our interest in freely expressing our moral and political opinions—and this interest sets internal parameters around what we take to be reasonable and unreasonable assertions of that right, or what types of speech-acts it is meant to protect. In this sense, we might say that the right to freedom of expression protects one's ability to protest on the steps of a state legislature, but it does not necessarily protect one's ability to shout racial epithets on a public bus. By contrast, the right to freedom of expression may also be subject to external limitations that derive from competing moral considerations. For example, while protesting on the steps of a legislature is a legitimate exercise of the right to freedom of expression, a police force may nevertheless be justified in breaking up a demonstration if a crowd is incited to violence and presents a legitimate risk to public safety.

The parental interest justification imposes at least two sets of internal limitations on the right to procreate. First, if the right to procreate is ultimately grounded in the strong interest that adults have in developing intimate fiduciary relationships with children, then it will *only* protect instances of procreation that are undertaken with the intention of parenting the resultant child and where the procreator has the ability to do so to an adequate standard. To understand why this is significant, consider two notable cases in which these conditions were not met.[63] The first is the case of David Oakley, a Wisconsin man who was convicted of the intentional failure to pay child support to the nine children he had fathered with four different women. As a result of his conviction, Oakley was placed on probation for a period of five years and subject to the condition that he not father any additional children until he could demonstrate the ability to support his existing children as well as any subsequent children he might create. The second is the case of "Ms. A," an Israeli woman who was denied IVF treatment based on concerns about her ability to parent the resultant child. Ms. A was experiencing significant psychological and social challenges at the time she sought IVF, and a pre-treatment report determined that her parenting capacity was sufficiently low to justify the refusal of treatment.

Both of these cases elicited public controversy on the grounds that their subject's right to procreate had been violated, though the parental interest justification does not support these judgments. Whatever else might be said about the appropriateness or constitutionality of Oakley's probation condition, it did not compromise his interest in parenting—indeed, if Oakley truly wanted to become a parent, he was free and in fact encouraged to do so with respect to any of the nine children he was already responsible for creating. The only interest that was compromised by Oakley's probation condition was his interest in the mere creation of biological children, though, as I have argued above, this is not the interest that ultimately justifies the right to procreate. There is no right to create children *simpliciter*, but only a right to create children with whom one can initiate parent-child relationships, and this right was not violated in Oakley's case.

The case of Ms. A is different in that, unlike Oakley, Ms. A *did* intend to parent the children she was seeking to create through IVF. However, she failed to meet a key eligibility criterion of the more fundamental right from which the right to procreate is derived. As philosophers like Brighouse and Swift have emphasized, the right to become and remain a parent is conditional on discharging one's fiduciary duties to an adequate standard. The inability to adequately discharge these duties makes one ineligible for the right to parent; and if one is ineligible for the right to parent, then it follows that one is ineligible for the right to procreate as a means of initiating a parent-child relationship.

Second, if the right to procreate is ultimately grounded in the strong interest that adults have in developing intimate fiduciary relationships with children, then it will not *necessarily* protect instances of procreation that are not necessary to realize this interest. Consider two cases in which this limitation might apply:

Restriction on Fertility Treatment: Jill wants to become a parent, but she is unable to conceive a child and requires the use of *in-vitro fertilization* (IVF) to initiate a pregnancy. When she approaches her local health authority to inquire about IVF treatment, she is informed of a new law that has recently come into effect that restricts the provision of fertility treatment in order to conserve limited health resources in the community. However, the health authority has partnered with a local adoption agency to connect IVF patients with children in need of parental care, and they inform Jill that there is currently a healthy newborn available for adoption. Jill refuses this offer and objects to the new law on the grounds that it violates her right to procreate.

Restriction on Family Size: Jane lives in a small island nation that is expected to be heavily impacted by the effects of climate change. In response to this threat, the government has undertaken a number of initiatives designed to reduce the country's total level of environmental impact and improve its sustainability. In addition to investments in green technology and enforceable limits on consumption, the government has also issued a non-enforceable recommendation that citizens limit the size of their families by restricting themselves to the creation of just one child. Jane, who would prefer to enjoy parent-child relationships with multiple children, objects to this recommendation on the grounds that it undermines her right to procreate.

The parental interest justification of the right to procreate does not offer a strong basis for either woman's complaint. Consider first Jill's complaint. If Jill has a legitimate complaint about the restriction on fertility treatment that is grounded in her right to procreate, then this must be because it precludes the satisfaction of the interest that is protected by that right. However, this is not the case on the parental interest justification. According to this account, the value of procreation is explained primarily as means to access the goods associated with parent-child relationships, and the value of parent-child relationships is conceived independently of any biological connection between parent and child—as Brighouse and Swift put it, everything that is valuable about parent-child relationships would still obtain "in a world in which children were produced by storks."[64] Thus, insofar as Jill is still able to satisfy this interest with respect to an infant who is available for adoption, she does not have a strong complaint about the restriction on fertility treatment (assuming there are sufficiently weighty moral reasons in favor of it). Of course, she might still have a complaint that is rooted in other types of interests—for example, under the restriction on fertility treatment, Jill is unable to realize her interest in experiencing pregnancy or in parenting a child who is biologically related to her—however, these interests are perhaps best characterized as discretionary interests that are not sufficiently important to justify the imposition of duties on others, and which accordingly carry less moral weight in the case under consideration.

Consider next Jane's complaint. If Jane has a legitimate complaint about the restriction on family size that is grounded in her right to procreate, then this must be because adhering to this restriction precludes the satisfaction of the interest that is protected by that right. But this again is not true: Jane is still able to satisfy her interest in parenting under the proposed restriction by creating and raising one child. Of course, a restriction that precluded the creation of any children full stop *would* give rise to such a complaint, as it would preclude the satisfaction of the interest the right to procreate is meant to protect. However, as the interest in parenting can be satisfied by initiating a parent-child relationship with just one child, this type of complaint does not emerge in *Restriction on Family Size*. This reveals an important aspect of the interest in parenting, which we might refer to as its *satiability*: once one has become a parent, their interest in parenting has been satisfied, and thus any interest they may have in parenting subsequent children will carry significantly less moral weight. This, of course, is not to suggest that the right to procreate only entitles its holder to procreate just once or that it cannot support the creation of multiple children with whom one can enjoy valuable parent-child relationships. However, it does suggest that in cases like *Restriction on Family Size*—where one's interest in parenting has been satisfied, and where there are independent moral considerations weighing against the creation of additional children—the right to procreate does not necessarily supply an objection to restrictions on procreation.

Of course, independent moral considerations can also impose external limitations on the right to procreate *even* in cases where procreation is necessary to realize an interest in parenting. This might be the case, for example, if the resultant child will be born into less than minimally decent conditions or if others' basic welfare interests are jeopardized by the introduction of new members into the community, including existing children in need of parental care or present and future people who are vulnerable to the environmental impacts of unconstrained population growth. A key task in the remainder of this book will be exploring these external limitations in greater detail and, in so doing, providing an account of the conditions under which it is morally permissible to exercise one's right to procreate.

Notes

1 For an excellent overview of selective reproductive technologies and the distinctive ethical issues they raise, see Stephen Wilkinson. *Choosing Tomorrow's Children: The Ethics of Selective Reproduction* (Oxford: Oxford University Press, 2011).
2 UN General Assembly. *Universal Declaration of Human Rights*. 10 December 1948, 217A(III), Article 16(1) and (3).
3 UN General Assembly. *United Nations International Conference on Population and Development Programme of Action*. 18 October 1994, A/CONF.171/13, paragraph 7.3.
4 UN General Assembly. *International Covenant on Civil and Political Rights*. 23 March 1976, 2200A (XXI), Article 23(2). ("The right of men and woman of marriageable age to marry and found a family shall be recognized.")

5 Council of Europe. *European Convention on Human Rights*. 3 September 1953, Article 12. ("Men and woman of marriageable age have a right to marry and found a family, according to the national laws governing the exercise of this right.")

6 Some notable exceptions include Daniela Cutas. "Sex is Overrated: On the Right to Reproduce," *Human Fertility*, 12(1) (2009), pp. 45–52; Luara Ferracioli. "On the Human Right to Found a Family," in Jesse Tomalty and Kerri Woods (eds.) *The Routledge Handbook for the Philosophy of Human Rights* (New York: Routledge, forthcoming); Tim Meijers. "The Value in Procreation: A Pro-Tanto Case for a Limited and Conditional Right to Procreate," *Journal of Value Inquiry*, 54(4) (2020), pp. 627–647; Yvette E. Pearson. "Storks, Cabbage Patches, and the Right to Procreate," *Journal of Bioethical Inquiry*, 4(2) (2007), pp. 105–115; Muireann Quigley. "A Right to Reproduce?" *Bioethics*, 24(8) (2010), pp. 403–411; and Daniel Statman. "The Right to Parenthood: An Argument for a Narrow Interpretation," *Ethical Perspectives*, 10(3–4) (2003), pp. 224–235.

7 There is an ongoing debate in philosophy regarding the nature and function of rights, principally between proponents of the *will theory* and the *interest theory*. However, as wading into this debate would take us too far afield, I will simply assume without argument in this book that rights are best understood as morally protected interests. For an excellent overview of the debate between will theorists and interest theorists, see Matthew Kramer, N.E. Simmonds, and Hillel Steiner. *A Debate over Rights: Philosophical Enquiries* (Oxford: Oxford University Press, 1998).

8 Wesley Hohfeld. *Fundamental Legal Conceptions*. Ed. W.W. Cook (New Haven: Yale University Press, 1919), 36.

9 For contemporary statements of the interest theory of rights, see Joseph Raz. *The Morality of Freedom* (Oxford: Clarendon Press, 1986), chapters 7 and 8; Matthew Kramer, "Rights without Trimmings," in Kramer, Simmonds, and Steiner, *A Debate over Rights*; and Neil MacCormick. "Rights in Legislation," in P.M.S. Hacker and Joseph Raz (eds.) *Law, Morality, and Society: Essays in Honour of H.L.A. Hart* (Oxford: Clarendon Press, 1977).

10 Raz, 1986, 166.

11 L.W. Sumner. *The Moral Foundations of Rights* (New York: Oxford University Press, 1987), 15.

12 Joel Feinberg. *Harm to Others, Volume 1 of the Moral Limits of the Criminal Law* (New York: Oxford University Press, 1984), chapter 2.

13 This is not to say that an interest in sexual intimacy could not still ground a negative right to non-interference in one's pursuit of sexual intimacy. This type of right has been—and continues to be—crucially important, particularly for sexual minorities who face credible threats of coercion or violence as a result of their pursuit of sexual intimacy.

14 Ronald Dworkin. "Rights as Trumps," in Jeremy Waldron (ed.) *Theories of Rights* (Oxford: Oxford University Press, 1984).

15 Robert Nozick. *Anarchy, State, and Utopia*. New York: Basic Books, 1974.

16 Jeremy Waldron. "Rights in Conflict," *Ethics*, 99(3) (1989), pp. 503–519, esp. 513–515.

17 Waldron, "Rights in Conflict," 513. For Rawls' original discussion of the idea of lexical priority, see John Rawls. *A Theory of Justice* (Cambridge, MA: Harvard University Press, 1971), pp. 42–44.

18 See, for example, Hillel Steiner. *An Essay on Rights* (Oxford: Blackwell, 1994).

19 John A. Robertson. *Children of Choice: Freedom and the New Reproductive Technologies* (Princeton: Princeton University Press, 1994), 24.
20 John Harris. "The Right to Found a Family" in Geoffrey Scarre (ed.) *Children, Parents, and Politics* (Cambridge: Cambridge University Press, 1989), 133.
21 Raz, *The Morality of Freedom*, 168–170.
22 U. Van den Broeke et al. "A Systematic Review of Sperm Donors: Demographic Characteristics, Attitudes, Motives, and Experiences of the Process of Sperm Donation," *Human Reproduction Update*, 19(1) (2013), pp. 37–51, 44.
23 S. Matthew Liao. *The Right to Be Loved* (Oxford: Oxford University Press, 2015), 155–156.
24 Elizabeth Brake. "Creation Theory: Do Genetic Ties Matter?" in Sarah Hannan, Samantha Brennan, and Richard Vernon (eds.) *Permissible Progeny? The Morality of Procreation and Parenting* (New York: Oxford University Press, 2015).
25 For an account of pregnancy as an epistemically transformative experience, see Fiona Woolard. "Mother Knows Best: Pregnancy, Applied Ethics, and Epistemically Transformative Experiences," *Journal of Applied Philosophy*, 38(1) (2021), pp. 155–171. For a more general account of the phenomenology of pregnancy, see Amy Mullin. *Reconceiving Pregnancy and Childcare: Ethics, Experience, and Reproductive Labor* (Cambridge: Cambridge University Press, 2005).
26 Tina Rulli. "Preferring a Genetically-Related Child," *Journal of Moral Philosophy*, 13(6) (2016), pp. 669–698, 696.
27 Plato. *Republic*. Translated by G.M.A. Grube, revised by C.D.C. Reeve (Indianapolis: Hackett, 1992), Book VI.
28 Christine Overall. *Why Have Children? The Ethical Debate* (Cambridge, MA: MIT Press, 2012), 21. Overall concedes that "a similar claim might be made about men with regard the use of men's gametes but the fact that gestation is a female condition makes respect for bodily freedom and autonomy particularly significant for women."
29 Laura Shanner. "The Right to Procreate: Where Rights Claims Have Gone Wrong," *McGill Law Journal*, 40 (1995), pp. 823–874, 841.
30 Overall, *Why Have Children?*, 26–27.
31 Dan W. Brock. "Shaping Future Children: Parental Rights and Societal Interests," *Journal of Political Philosophy*, 13(4) (2005), pp. 377–398, 382. For a similar account, see also Allen Buchanan, Dan W. Brock, Norman Daniels, and Daniel Wikler. *From Chance to Choice: Genetics and Justice* (Cambridge: Cambridge University Press, 2000), chapter 6.
32 Brock, "Shaping Future Children," 380.
33 This, of course, is not to suggest that these two types of interference are morally equivalent, but simply that they both restrict a person's ability to carry out a particular conception of the good. There are additional (including bodily-integrity-based) objections to coercive sterilization that do not apply to restrictions on one's access to gametes or fertility treatment.
34 Brock, "Shaping Future Children," 380.
35 Harry Brighouse and Adam Swift. *Family Values: The Ethics of Parent-Child Relationships* (Princeton: Princeton University Press, 2014), esp. chapter 4.
36 Brighouse and Swift, *Family Values*, 88.
37 Harry Brighouse and Adam Swift, "Parents' Rights and the Value of the Family," *Ethics*, 117(1) (2006), pp. 80–108, 93.
38 Brighouse and Swift, "Parents' Rights and the Value of the Family," 95.
39 Brighouse and Swift, *Family Values*, 93–97.

40 For a compelling defense of a child's interest in continuity of care, see Anne L. Alstott. *No Exit: What Parents Owe Their Children and What Society Owes Parents* (Oxford: Oxford University Press, 2004), esp. chapters 1 and 2.
41 For Brighouse and Swift, this reveals that the right to parent must be grounded at least partially in parents'—and not just children's—interests. I discuss different accounts of the justification of parental rights in Chapter 3, Section 3.3.
42 Brighouse and Swift, *Family Values*, 55.
43 Brighouse and Swift, *Family Values*, 178.
44 I will discuss the costs associated with adoption and their implications for the morality of procreation in Chapter 6, particularly Section 6.4.
45 Sarah Hannan and R.J. Leland. "Childhood Bads, Parenting Goods, and the Right to Procreate," *Critical Review of International Social and Political Philosophy*, 21(3) (2018), pp. 366–384, 367.
46 Hannan and Leland. "Childhood Bads, Parenting Goods, and the Right to Procreate," 370. For a further elaboration and defense of the purportedly bad-making features of childhood, see also Sarah Hannan. "Why Childhood is Bad for Children," *Journal of Applied Philosophy*, 35(S1) (2018), pp. 11–28.
47 Hannan and Leland, "Childhood Bads," 377.
48 Hannan and Leland, "Childhood Bads," 367.
49 See, for example, Brighouse and Swift, *Family Values*, 62–65; Samantha Brennan. "The Goods of Childhood and Children's Rights," in Françoise Baylis and Carolyn McLeod (eds.) *Family-Making: Contemporary Ethical Challenges* (Oxford: Oxford University Press, 2014), and Anca Gheaus. "Unfinished Adults and Defective Children: On the Nature and Value of Childhood," *Journal of Ethics and Social Philosophy*, 9(1) (2015), pp. 1–21.
50 Brennan, "The Goods of Childhood," 37.
51 Some proponents of anti-natalism would deny that there are any such benefits, and hence that procreation is always an impermissible means of initiating a parent-child relationship. If this position is sound, then the parental interest justification fails, though as I indicated in the Introduction, I am bracketing discussion of anti-natalism from this book and assuming that procreation is at least presumptively morally impermissible. For an account of anti-natalism that denies existence is ever beneficial, see David Benatar. *Better Never to Have Been: The Harm of Coming into Existence* (Oxford: Oxford University Press, 2006).
52 See Anca Gheaus. "Biological Parenthood: Gestational, Not Genetic," *Australasian Journal of Philosophy*, 96(2) (2018), pp. 225–240, and "The Best Available Parent," *Ethics*, 131(3) (2021), pp. 431–459.
53 See J. David Velleman. "Family History," *Philosophical Papers*, 34(3) (2005), pp. 357–378, and Bernard Prusak. *Parental Obligations and Bioethics: The Duties of a Creator* (New York: Routledge, 2013), esp. chapter 1.
54 Luara Ferracioli. "Procreative-Parenting, Love's Reasons, and the Demands of Morality," *Philosophical Quarterly*, 68(270) (2018), pp. 77–97.
55 Ferracioli, "Procreative-Parenting, Love's Reasons, and the Demands of Morality," 90.
56 Ferracioli, "Procreative-Parenting, Love's Reasons, and the Demands of Morality," 88.
57 Ferracioli, "Procreative-Parenting, Love's Reasons, and the Demands of Morality," 88–89.
58 Sam Shpall raises a similar objection in an insightful multi-pronged critique of Ferracioli's view. See "Parental Love and Procreation," *Philosophical Quarterly*, 73(1) (2023), pp. 206–226, esp. 210–211.

59 Overall, *Why Have Children?*, 22.
60 Robertson, *Children of Choice*, 29.
61 Robertson, *Children of Choice*, 26.
62 This distinction and its application to the right to procreate is drawn from David Archard. "Wrongful Life," *Philosophy*, 79 (309) (2004), pp. 403–420, 414–416.
63 These cases are drawn from Statman, "The Right to Parenthood," 224–225.
64 Brighouse and Swift, *Family Values*, 104.

2 The Child's Right to a Decent Start

2.1 Introduction

In the previous chapter, I argued that a strong interest in accessing the goods associated with parent-child relationships can ground a moral right to procreate, albeit one that is limited and conditional. The right to procreate is limited in the sense that it only protects instances of procreation in which there is a genuine parental interest at stake. As such, it does not protect instances of procreation in which there is no intention or ability to parent the resultant child, and it does not necessarily protect instances of procreation that are not strictly necessary to realize an interest in parenting (e.g., the creation of multiple children). The right to procreate is also conditional in the sense that it only protects instances of procreation that do not infringe on the legitimate moral entitlements of other people, particularly the class of basic welfare interests that can be understood to have lexical priority over the project-based interest in parenting. Thus, a central task in developing a theory of procreative justice is to determine whose moral entitlements may be impacted by one's procreative decisions, and which of these moral entitlements have lexical priority over the interest in parenting.

I begin this task in the present chapter by considering how the interests of prospective children constrain the moral permissibility of procreation. Given the profound impact of procreation on the welfare and life prospects of the children who result from it, it is reasonable to think that the right to procreate ought to be constrained in light of their interests, perhaps only permitting the creation of children who can reasonably be expected to live (at the very least) minimally decent lives. This is consistent with ordinary understandings of the scope and limits of other parenthood-related rights, for example, the acquisition and maintenance of which is normally conditional on advancing the child's interests to some minimum degree. However, there is a philosophical problem in the case of procreation that complicates this reasoning and threatens to limit the extent to which children's interests

DOI: 10.4324/9781003164104-3

can be incorporated into a theory of procreative justice.[1] When a set of parents fail to advance their existing child's interests, there is an identifiable person who is made worse off by that failure—their child's life goes worse than it otherwise would have gone had they fulfilled their responsibilities and advanced the child's interests. However, the same is not true in the case of procreation. When a set of procreators bring their child into disadvantaged conditions, the child is not usually made any worse off than they otherwise would have been, as their very existence depended on that procreative act taking place how and when it did. In this sense, as Dan Brock has observed,

> actions with effects for a child that would constitute seriously wrongful child abuse if done to an existing child are no harm, and so no wrong, if done to a child when they are inextricable from the decision to bring that child into existence with a worthwhile life.[2]

In the existing literature surrounding procreative ethics, this type of reasoning has sometimes been used to exclude children's interests from the set of considerations that may be appealed to when evaluating the morality of procreation. On this view, either it is not wrong to bring a child into disadvantaged conditions because that child will not be made any worse off, or it is wrong to bring a child into disadvantaged conditions, but its wrongness must be explained in non-child-affecting terms (e.g., in terms of a failure to maximize total utility or in terms of a lack of sensitivity displayed toward the welfare of one's future offspring). In this chapter, I argue that this reasoning is flawed and that it is in fact possible to make sense of the idea that bringing a child into sufficiently disadvantaged conditions wrongs that child. Even if a child cannot be made comparatively worse off by being born into disadvantaged conditions, they may nevertheless be wronged if they are born into conditions that fall below a minimally decent welfare threshold to which all existing children are entitled. The wrong in question can be understood in terms of violating the child's *right to a decent start*, a right that functions to limit the scope of their would-be parents' right to procreate.

The chapter proceeds in four parts. In Section 2.2, I outline the non-identity problem and the central challenge that it poses in the context of procreative ethics, namely, that a child cannot be harmed by a procreative act insofar as their existence depended on that act taking place. In Section 2.3, I explain how an appeal to rights can get around this problem by accounting for the moral wrong in non-identity cases without insisting that the child has been made comparatively worse off. I also address the concern that an appeal to rights in this context is objectionably *ad hoc*. In Section 2.4, I outline and respond to five prominent objections that have been raised against rights-based solutions in the contemporary literature. Finally, in Section 2.5, I introduce the central normative implication of my argument—the *decent start principle*—and clarify its role within a broader account of procreative justice.

2.2 The Non-Identity Problem

The non-identity problem arises in cases where an act that apparently harms a person is also a necessary condition for their existence. It is generated from the mundane fact that a person's numerical identity is tied to the timing of conception, such that if your parents had not conceived when they in fact did, the resultant child would not have been you, but a numerically distinct "non-identical" child composed of a unique set of reproductive gametes. Parfit expresses this fact as the *Time-Dependence Claim*, which holds that "If any particular person had not been conceived when he was in fact conceived, it is *in fact* true that he would never have existed."[3] According to Parfit and others, the Time-Dependence Claim forces us to concede that, so long as a person has a life worth living, they cannot be harmed by a set of actions or choices that led to their conception, for their very existence depended on that conception taking place when it did.

The full force of the non-identity problem is felt in cases where our actions or choices cause a person to be born into disadvantaged conditions. While our gut reaction may be to claim that such actions or choices are wrong—and wrong in light of the adverse effects on the person who results—the non-identity problem complicates this inference. Because a person's numerical identity is tied to the timing of conception, they often owe their existence to the very act that appears to harm them—for had that act not been performed, they may never have been conceived, and in this sense are not made any worse off than they otherwise would have been. The logic of the non-identity problem is best illustrated through a series of hypothetical cases designed to show that an apparently harmful act is not in fact worse for anyone. As a paradigm non-identity case, consider Parfit's *14-Year-Old Girl*:

> *14-Year-Old Girl*: This girl chooses to have a child. Because she is so young, she gives her a child a bad start in life. Though this will have bad effects throughout the child's life, his life will, predictably, be worth living. If this girl had waited for several years, she would have had a different child, to whom she would have given a better start in life.[4]

What is puzzling about *14-Year-Old Girl* and other non-identity cases is that there seems to be a clear moral wrong that has taken place, yet no identifiable person who is made any worse off than they otherwise would have been. Due to Time-Dependence, the circumstances that the girl's child was born into were the best that *he* could have hoped for. Had the girl waited until she was older and more mature, the resultant child would not have been him, but rather a different child altogether, one composed of a unique set of reproductive gametes. Assuming that his disadvantaged life is preferable to no life at all—even marginally—he has not been made any worse off than he otherwise would have been. The challenge, then, is in articulating what, if anything, was wrong with his mother's decision.

The non-identity problem is often presented as a fact that poses an insurmountable challenge to claims of wrongdoing to future persons, though it actually relies on three contestable assumptions, the combination of which generates a paradox:

1 the act in question is morally objectionable;
2 for an act to be morally objectionable, it must wrong a particular person; and
3 for an act to wrong a particular person, it must make them worse off.

The puzzle of non-identity cases arises from the fact that, while each of these assumptions is intuitively appealing, they cannot all be held at once given that no one has been made worse off. Thus, responses to the non-identity problem have mostly centered along three lines, with each challenging or dispensing with a particular assumption.

First, some theorists have dispensed with (1) and simply conceded that the act in question is not in fact morally objectionable.[5] This set of views retains assumptions (2) and (3) and holds that a morally objectionable act must make a particular person worse off than they otherwise would have been. On the basis of these assumptions, it deduces that because a particular person did not exist at the time of the act in question, and would not have existed had the act not been performed, they cannot be made worse off by the act, and so the act must therefore be morally permissible. In its strongest guise, this set of views yields the blunt conclusion that future persons who depend on our choices for their existence have "no moral status of any kind, not even a weak one."[6]

Second, some theorists have dispensed with (2) and insisted that an act can be morally objectionable without wronging any particular person.[7] According to this set of views, the moral failing in non-identity cases does not consist in violating a duty to any particular person, but rather in violating a duty to promote certain impersonal values, such as utility or beneficence. On this view, the moral wrong in cases like *14-Year-Old Girl* does not consist in causing harm to the particular child who results, but rather in failing to create a better off child in his place, or in "wantonly introducing a certain evil into the world."[8]

Finally, some theorists have dispensed with (3) and have argued that an act can wrong a particular person without making them worse off. There are several variations on this line of response. First, some theorists have denied that a person must be made worse off in order to be harmed—that is, they have rejected a comparative conception of harm in favor of a non-comparative conception.[9] On this understanding, harm does not consist in being made worse off than one otherwise would have been, but rather in being subjected to non-comparatively harmful states, such as "pain, early death, bodily damage, or deformity," among others.[10] Adopting a non-comparative conception of harm allows theorists to claim that the moral wrong in non-identity cases *does* consist in causing harm to the particular person who results, even if that person is not made worse off than they otherwise would have been.

Alternatively, other theorists have distinguished between harm and wrongdoing and have argued that even if a person cannot be comparatively harmed in non-identity cases, they may nevertheless be wronged. One set of views locates the moral wrong purely in the actions or attitudes of the agents in non-identity cases, e.g., through their lack of concern or respect for the person born into disadvantaged conditions.[11] Another set of views locates the moral wrong in the not-worse-off consequences for the apparent victim in non-identity cases, e.g., in their being born into conditions that fall below a minimally decent welfare threshold to which all persons are putatively entitled.[12]

My objective in this chapter is to defend the tenability of one variation of this third line of response, namely, that the moral wrong in cases like *14-Year-Old Girl* can be plausibly understood in terms of violating the child's right to be born into minimally decent conditions. Successfully defending this solution is sufficient to show that the first line of response is false, though it does not show, nor is it intended to show, that any other line of response is false. As I will explain in further detail in Section 2.5, the rights-based view defended in this chapter is consistent with a pluralistic approach toward non-identity cases that takes a range of impersonal and person-affecting considerations into account when evaluating their putative moral wrongness. However, since my aim in this chapter is to explore if and how children may be wronged in a person-affecting sense when they are born into disadvantaged conditions, I will set aside discussion of alternative solutions and focus my attention on defending what I take to be a plausible rights-based solution.

2.3 A Rights-Based Solution

How, then, does such a solution work? As I noted previously, all rights-based solutions begin with the more general premise that, contrary to assumption (3), a person can be wronged by an act without being made comparatively worse off by it. This is a crucial premise, as it allows for the possibility that a child may be wronged by being born into disadvantaged conditions even if they are not made any worse off than they otherwise would have been. Many theorists have taken this premise to be clearly false, claiming that if what we are doing will not be worse for a particular person, or will even be better for a particular person, then we are not in a morally relevant sense wronging that person.[13] However, as James Woodward and others have noted, there are some instances in which this premise seems to be clearly true, including basic instances of rights violations.[14] Consider, for example, Woodward's case of *Smith*:

> *Smith*: Suppose that Smith, who is black, attempts to buy a ticket on a certain airline flight, and that airline refuses to sell it to him because it discriminates racially. Shortly after, that very flight crashes, killing all aboard.[15]

Clearly the airline's refusal has left Smith, on balance, better off than he otherwise would have been, for had he boarded the aircraft, he would have died along with the other passengers. However, even though Smith has not been made comparatively worse off by the airline's refusal, he has nevertheless been wronged by being stigmatized and treated unequally on the basis of his race. His right to non-discrimination has been violated, and he retains a legitimate complaint against the airline *even if* its refusal left him better off than he otherwise would have been.

More familiar examples arrive at the same point. Factories operating in developing countries often exploit impoverished locals by employing them in low-paying jobs amid dangerous working conditions. In some of these cases, the workers are made better off than they otherwise would have been, as there are no more lucrative opportunities for work. However, even if the workers are made better off on balance, it is entirely consistent to say that they are nevertheless wronged by receiving less-than-fair compensation for their labor. This points to the more general fact that an act can wrong a person independently of its effect on their overall welfare. It can also wrong them for non-welfarist reasons as well, e.g., for violating their rights.

Rights-based solutions appeal to this fact in order to account for the moral wrong in (some)[16] non-identity cases. Even if a procreative act cannot make a child comparatively worse off, it still may be wrong by violating one or more of that child's rights, such as a putative right against being born into a certain kind of existence. On this view, what is potentially wrong with a case like *14-Year-Old Girl* is not that the child is made any worse off than he otherwise would have been, but rather that he is made worse off than he is *entitled to be* according to a plausible conception of children's rights: all children arguably have a right to be born into conditions in which they will enjoy minimally decent parental care, and this right was violated by his mother's procreative act. A rights-based solution thus has the advantage of accounting for the moral wrong in non-identity cases while preserving the intuition that it is the *resultant child* who has been wronged. In other words, it is able to account for the moral wrong in child-affecting terms.

As specified so far, however, a rights-based solution offers only a formal response to non-identity cases, and the extent to which it impugns certain procreative acts will depend on the substantive content of a child's rights. Within the literature, this content differs from theorist to theorist and is articulated with varying degrees of specificity. For instance, Bonnie Steinbock claims that every child has a right to a life that "holds a reasonable promise of containing the things that make human lives good," such as "an ability to experience pleasure, to learn, [and] to have relationships with others."[17] Alternatively, David Archard claims that a child has a right to a life in which they can enjoy a significant number of the rights articulated within the United Nations Convention on the Rights of the Child (1989), which include but extend far beyond the minimal criteria presented by Steinbock.[18] In this sense, while all rights-based solutions include what Archard calls a "birth-right claim," or the notion that all children have a right to a standard of living above a certain threshold, exactly *where* that threshold is set is a matter of reasonable disagreement.

In order to avoid the charge of arbitrariness, I believe this threshold should be informed by the most defensible account of what we owe to children once they exist. Thus, in what follows, I will posit a sufficientarian threshold that focuses on the satisfaction of children's basic welfare interests, or the interests that must be satisfied for them to lead minimally decent lives as children and to grow into independent and functional adults. This modest conception of children's rights enjoys widespread support in the existing literature on the morality of parenting and can be understood to follow straightforwardly from the interest theory of rights, according to which "X has a right if (and only if) an aspect of X's well-being (his interest) is a sufficient reason for holding some other person(s) to be under a duty."[19] At a minimum, I take the relevant interests to include an interest in adequate health, nutrition, shelter, protection from neglect and abuse, and access to adequate parental care, though others may be reasonably added to this list. On this view, what all children have a right to is to be born into conditions in which their basic welfare interests can be satisfied for the duration of their childhood, which I will refer to as the child's *right to a decent start*. Accordingly, intentionally creating a child amid conditions in which those interests cannot be satisfied violates that child's right and constitutes a wrong to that child.

I will address five prominent objections to this view in the following section, though before doing so, it is first necessary to sideline an important objection to the idea of positing a rights threshold in the first place, particularly one that is set above the more minimal standard of a "life worth living." One common worry about rights-based solutions in general is that they can seem suspiciously *ad hoc*: while appealing to a child's right to a decent start might allow us to account for the moral wrong in (some) non-identity cases, it is not always clear that we have a reason for attributing such a right to children independently of the fact that it allows us to account for that moral wrong.[20] Moreover, critics who advance this worry typically point out that we *do* seem to have independent reason for attributing to children a more minimal right to a "life worth living," for if their lives are *not* worth living—in the sense that non-existence would be rationally preferable—then they can coherently claim to have been made worse off by being brought into existence.[21]

Do these considerations undermine the sufficientarian view outlined above? Not necessarily, for two reasons. First, it seems clear that we *do* have independent reason for attributing to children a right to be born into conditions in which their basic welfare interests can be satisfied for the duration of their childhood. This right, attributed to all future children, is simply a corollary of the rights that all existent children will have once they are born. Thus, to the extent that we have independent reason for attributing *x* rights to existent children, we also have independent reason for attributing to future children a right to be born into conditions in which *x* rights can be fulfilled.[22] We can reasonably disagree about what these rights are, though it is plausible to think they include (at the very least) rights to the satisfaction of the basic interests children require to lead minimally decent lives.

Second, once we separate the notions of harming and wronging in the way that a rights-based solution suggests, the more minimal threshold of a "life worth living" begins to look *ad hoc* itself. Standard analysis of the non-identity problem typically includes an exception for cases involving lives not worth living, for in these types of cases, we can coherently claim that a person has been made worse off by being brought into existence. However, if being wronged is not coextensive with being harmed, then it is not clear why comparative worsening should be a necessary condition of wrongdoing in procreative cases. The mere fact that a child has not been made worse off by being brought into existence does not tell us anything about whether she has been wronged unless we assume that children are *only* entitled to lives worth living.[23] However, it is unclear what independent reason we have to make *this* assumption. In addition to negative rights not to be harmed in various ways, we normally take children to have positive rights to the provision of basic goods and opportunities like those listed above. To the extent that this is correct, a sufficientarian threshold based on children's basic welfare interests is *more* independently justifiable than a lower threshold based on a life worth living, as it is rooted in a more familiar and ultimately more plausible conception of children's rights.

2.4 Objections

Even if we drop assumption (3), and insist that a child may be wronged by being born into disadvantaged conditions without being made comparatively worse off by it, vestiges of the non-identity problem remain insofar as that child's numerical identity is tied to the timing of conception. Parfit points this out by claiming that a child who has his rights violated through a procreative act could not have had *his* rights fulfilled, for his parents could not have had *him* by conceiving under more favorable conditions, but only a better off, different child.[24] If ought implies can, the fact of non-identity might appear to get his parents off the moral hook. After all, if his right to a decent start could not possibly have been fulfilled, in what sense could his parents have violated it?

One way of responding to this point is to simply reframe the moral wrong in terms of creating a child whose right cannot be fulfilled, something Woodward likens to making a promise you know you cannot keep.[25] However, even this qualified view is open to a number of possible objections, each of which challenges the invocation of rights in procreative contexts.

2.4.1 *The Identifiability Objection*

The first objection insists that future children do not satisfy a necessary condition of rights-ascription and therefore cannot be said to have their rights violated through a procreative act. David Heyd notes that in our usual understanding, rights seem to be linked to metaphysically identifiable people who are the subjects of those rights and the recipients of their correlative duties.[26] For instance, in the case of *Smith*, we are able to pick out a particular person who is the subject of a right not to be discriminated against, and thus a particular person to whom a duty of non-discrimination is owed. It then makes sense to say that Smith's right was violated because the airline failed to perform a duty owed specifically to Smith as an identifiable person.

Future children, however, are not identifiable in this way. As Michael Bayles points out, there are deep epistemic limitations to identifying the future children who we might create, as the designator "future child" may refer to *any* of the possible children we might create depending on which sperm fertilizes which egg.[27] In this sense, it is not clear to whom, exactly, our procreative duties are owed. But as Heyd makes clear, the problem of identifiability runs even deeper than this, for it is not only that the identities of future children are beyond human knowledge—they are also wholly dependent on human choice. Future children are not concrete individuals "waiting in the wings," so to speak, but depend for their identity on the human choices that bring about their conception. Ascribing rights to future children might then seem paradoxical, for if the identity of a future child depends on our choices, how can those same choices be constrained by consideration for *their* rights? Heyd concludes that they cannot be and that to apply rights in non-identity cases is to ignore their *sui generis* nature.

Is the identifiability objection sound? Perhaps partly. It is true that among the future children implicated by our procreative choices, there is no numerically identifiable person to whom we can ascribe a right, and thus no particular person to whom we can owe a correlative duty. But this fact is only normatively significant if we accept Heyd's assumption that rights must be linked to metaphysically identifiable people. This assumption seems false, however, and oddly out of touch with common understandings of how rights function. On one plausible understanding, rights do not attach to people's numerical identities, but rather to the *kinds* of beings they are, and the positions they occupy in relation to others. What is relevant to the ascription of rights, then, is not a person's numerical identity, but rather their relevant type-description, or the normatively significant set of characteristics that indicates what kind of being they are, and what kinds of interests they have. To see this point more clearly, consider once again the case of *Smith*. It is true that the individual Smith had a right not to be discriminated against and that by refusing to sell him a ticket, the airline violated Smith's right. But Smith did not hold this right by virtue of being the individual *Smith*, i.e., Smith, born June 17, 1952, on the basis of a prior conception that brought together a particular combination of Mr. and Mrs. Smith's reproductive gametes. Rather, Smith held this right on the basis of particular characteristics indicative of his *type*, i.e., a citizen who stands in a particular relation of equality to others. This right is also held by Jones, Brown, Wilson, and all the other citizens who share those particular characteristics, and the airline would have violated their rights in the same way by refusing to sell them tickets on the basis of their race.

The non-identifiability of future children should not prevent us from thinking about them as rights-holders to whom particular duties are owed, for we still know the *kinds* of beings they will become, and thus the kinds of interests they are likely to have. Regardless of their numerical identities, all children have certain interests *qua* children, the satisfaction of which is necessary for them to lead minimally decent lives. Thus, our procreative duties are owed to *all* the future children who we create, *whoever* they turn out to be. To this end, their numerical identities are not morally important.

2.4.2 *The Non-existence Objection*

Even if we pursue this line of reasoning and regard future children as a class of persons with antecedently identifiable interests, their eligibility as right-holders might be attacked from a different angle. In the existing literature surrounding intergenerational justice, many commentators have denied that rights can be ascribed *now* to members of future generations who do not currently exist.[28] The thought behind this claim is that the possession of rights necessarily presupposes the existence of the being who possesses them, such that "the ascription of rights is [only] made to actual persons—not possible persons."[29] On this view, members of future generations will come to have rights when they exist, but they cannot currently possess rights that impose correlative obligations on members of the present.

A similar objection might be raised in the smaller-scale context of procreation. While future children will come to have rights when they exist, their present non-existence might be thought to imply that they cannot currently possess rights that impose constraints on our procreative behavior. At the very least, any claim to the contrary would seem to rely on the controversial metaphysical claim that non-existent entities can possess real properties, and to the extent that we reject such a claim, we may be led to reject the notion that future children possess rights.

Is the non-existence objection sound? Again, perhaps partly. There is a coherent argument for the claim that future persons cannot possess rights *now*, expressed in the following set of premises:

1 in order to possess rights, one must have interests;
2 in order to have interests, one must exist;
3 future persons do not presently exist; therefore,
4 future persons cannot presently possess rights.

However, the fact that future persons cannot possess rights *now* does not preclude the possibility that we can have present obligations that are grounded in rights they *will* possess in the future. As an example of a case that illustrates this possibility, consider Joel Feinberg's kindergarten bomber:

> Imagine if you can a criminal so wicked that he wishes to blow up a schoolhouse to kill or mutilate pupils. He conceals a powerful bomb in the closet of the kindergarten room and sets a timing device to go off in six years' time. Six years later the bomb explodes, killing or mutilating dozens of five-year-old children.[30]

The five-year-old victims in Feinberg's example did not exist, and hence did not possess rights, at the time the bomb was planted, though this fact alone does not seem to preclude a rights-based objection to the criminal's act. At the time of his act, he knew (or at least had reason to expect) that there would be children present when the bomb exploded and that these children would *then* possess rights not to be killed or mutilated. Thus, there is nothing incoherent about claiming that the criminal had an obligation, grounded in the future rights of future children, to refrain from planting the bomb.

The children who we bring into existence through our procreative choices can similarly be thought of as future right-holders: *when* they come into existence, they will *then* have rights to the satisfaction of certain basic interests and will therefore be wronged when they are born into conditions in which those interests cannot be satisfied. Thus, it does not matter that the child of the *14-Year-Old Girl* does not exist at the time of her procreative decision, for we can still frame an objection to her decision that appeals to the rights he *will* possess upon coming into existence.

One might object that the preceding response to the non-existence objection is actually a red herring, for the children in non-identity cases are not (yet) *future* children who will exist in the future, but are rather *possible* children who may or may not exist in the future depending on the outcome of a particular choice. Thus, we cannot appeal to the rights of a child who *will* exist in the future in order to guide that choice, as the choice will in fact determine whether such a child ever exists. However, if we cannot appeal to the rights of a child who *will* exist in the future in order to guide such a choice, then it is not immediately clear how the preceding response to the non-existence objection is relevant for non-identity cases. Indeed, while we may have present obligations that are grounded in the rights of persons who *will* exist in the future, it does not necessarily follow that we have present obligations that are grounded in the rights of persons who *might* exist in the future.

There is a simple response to this objection that draws upon the previous response to the non-existence objection. The possible children in non-identity cases are possible future persons and can therefore be thought of as *contingent* future right-holders: *if* they are brought into existence, they will then have rights to the satisfaction of certain basic interests and will therefore be wronged if they are born into conditions in which those interests cannot be satisfied. Thus, when we refrain from creating such a person, we are not doing so out of concern for a merely possible person who is now spared a sub-threshold existence, but rather out of concern for the actual person who *would have* been born into sub-threshold conditions had we created them. Notice, however, that in cases where children *are* in fact born into sub-threshold conditions, we still need a response to the potential objection, raised *ex post*, that they could not have had their rights violated because they did not exist at the time of the act that resulted in the violation of their rights. The original response to the non-existence objection is therefore not a red herring, but rather a necessary component of our *ex post* judgment that the decision to create such a child was morally wrong.

2.4.3 *The Waiving of Rights Objection*

Suppose, then, that it is intelligible to think of future children as future rights-holders and that their rights may be violated when they are born into conditions in which their basic welfare interests cannot be satisfied. A further problem arises from the fact that such children may not regret the violation of the rights insofar as their existence depended on it occurring. Parfit makes this point by citing a letter from *The Times*, written in response to a British politician who welcomed the fact that there had been fewer teenage pregnancies than the previous year.[31] The man who wrote the letter had been born to a 14-year-old mother, and while he *did* have a difficult upbringing, he now enjoyed his life and balked at the suggestion that it would have been better had he never existed. According to Parfit, this man's reaction undermines rights-based objections to his mother's procreative conduct, for even if we claim that his mother violated his rights by bringing him into disadvantaged conditions, he might then choose to *waive* those rights to the extent that he is happy to be alive. The possibility of waiving rights might seem to diminish the significance of rights violations in procreative contexts. After all, in what sense is a rights violation objectionable if it is not regretted, and is even celebrated, by the supposed victim?

There are two different ways that we might go about responding to the waiving of rights objection. The first is to challenge the significance of a child's lack of regret by pointing to an intuitive asymmetry between the benchmarks associated with starting and continuing lives. According to some philosophers, we tend to assign a higher benchmark to starting lives than we do to continuing them, such that we may judge our lives to be worth continuing with certain disadvantages, but also think that it is better not to start a life with the same disadvantages.[32] For example, David Benatar notes that

> while most people think that living life without a limb does not make life so bad that it is worth ending, most (of the same) people also think that it is better not to bring into existence somebody who will lack a limb.[33]

This asymmetry might be thought to undermine the significance of a child's lack of regret, for the fact that a child (or the adult they become) judges their life to be worth continuing does not necessarily tell us anything about whether it was permissible to initiate in the first place. David Velleman appears to draw this conclusion, claiming that "we can hardly justify initiating a life on the mere grounds that there would not immediately be reason to terminate it."[34]

Of course, this response depends on an intuitive judgment that is not universally shared and thus may be open to an easy counter-response: perhaps the fact that life would still be worth continuing after losing a limb simply indicates that starting a life without a limb would not be so bad.[35] And if we take the benchmarks associated with starting and continuing lives to be symmetrical in this way, we might then consider a child's lack of regret to be centrally important, as the fact that a child judges his life to be worth continuing might suggest that it was permissible to initiate in the first place, despite the fact that his rights have been violated.

A second and more promising way of responding to the waiving of rights objection is to challenge its underlying principle of consent. Notice that the waiving of rights objection turns crucially on a principle of consent, or the idea that right-violating conduct can be rendered permissible by receiving the free consent of the person whom it affects. This idea is uncontroversial in cases where the person in question is available to offer their consent. For instance, while my right to bodily integrity prevents you from permissibly cutting me with a scalpel, I might choose to waive that right by allowing you to perform surgery on my heart. Assuming I was not coerced, deceived, and had adequate knowledge of the relevant facts and risks, I have meaningfully consented to the surgery, and you do no wrong by cutting me.

In the case of procreation, however, we cannot obtain the consent of the child prior to the act that violates their rights, for the act that violates their rights is also a necessary condition of their existence. Does this fact alone undermine the waiving of rights objection? Not necessarily, for as some of its proponents have pointed out, it is not always necessary to secure a person's consent prior to engaging in a presumptively right-violating act. Rather, it is sometimes sufficient that the person would *hypothetically* consent were they able to do so at the time of the act or that they will *subsequently* consent at a later time.[36] For example, suppose that I am unconscious because of my heart condition and you perform life-saving surgery on me without first obtaining my consent. In this case, it is true that I have not consented to the surgery, though it still seems permissible for you to perform it given the strong likelihood that (a) I *would have* consented to it had I been able to do so at the time of the surgery, or (b) I *will* consent to it later upon regaining consciousness. According to David Boonin, this type of example provides a strong basis for the waiving of rights objection, for even if a child cannot consent to being born into subthreshold conditions, it is reasonable to think either that (a) they *would have* consented to it had they been able to do so, or (b) they *will* consent to it at a later time.[37]

Do either of these possibilities undermine the proposed rights-based solution? Consider first the possibility of hypothetical consent. Initially, it might seem reasonable to think that if they were given the opportunity and made aware of the available options, a child would choose to waive their right to a decent start rather than to never exist at all. However, there are at least two problems with this type of reasoning. First, and most importantly, the case of procreation is disanalogous to the case of surgery in a way that makes an appeal to hypothetical consent problematic. When we infer hypothetical consent in the case of the unconscious patient, we are assuming they would have consented to the surgery at the moment it was performed. This is a reasonable assumption in most cases, as people typically have an interest in continuing to exist and would thereby be deprived by ceasing to exist. However, this type of assumption cannot play the same justificatory role in cases of procreation, for prior to the moment a child is created, there is no existent person who might have given their consent to being born into sub-threshold conditions—rather, there is only a possible person at that moment, one who will be wronged if they are born into sub-threshold conditions, but who will not otherwise be deprived if they are never created. An appeal to hypothetical consent is therefore incoherent, for it falsely attributes to possible people an interest they can only have once they exist.

However, even if hypothetical consent *could* be appealed to in the way that Boonin describes, a second problem is that the circumstances seem like those in which the inferred consent would not be morally binding. Suppose that it makes sense for us to imagine a possible person prior to existence who is faced with the choice of either existing in sub-threshold conditions or never existing at all. Even if it were intelligible to assume that such a person has an interest in existence, and would thereby choose the former over the latter option, it is not clear that this has the normative relevance Boonin ascribes to it, as their situation resembles one in which consent is obtained under conditions of duress.[38] If I hold a gun to your head and offer you a choice between your money or your life, it is reasonable to assume that you would agree to hand over your money. However, this type of agreement has no moral force, as consent that is obtained under the threat of violence is not a morally binding form of consent. By parity of reasoning, consent to a rights-violation that is obtained under the structurally similar threat of non-existence does not seem like a morally binding form of consent. Children in such a scenario face an option set that is itself unfair, so their hypothetical preference for the less bad option should not be taken as an exculpating factor in the violation of their rights.[39]

Taken together, these two problems severely undermine the hypothetical consent interpretation of the waiving of rights objection: not only is it incoherent to appeal to hypothetical consent in procreative contexts, but even if it *were* coherent, the inferred consent would be tainted in a way that renders it morally inert. But what about the subsequent consent interpretation, according to which children (or the adults they become) later consent to being born into sub-threshold conditions? This interpretation, reflected in Parfit's original example, might seem more promising given the fact that existing persons can and often do express an appreciation for having been born, even when they have been born into very difficult conditions. Can the expectation that a child will subsequently waive their right to a decent start justify the violation of that right?

I do not believe that it can, for a combination of reasons. First, and at a general level, it is not clear that an appeal to subsequent consent actually does any justificatory work in the analogous surgery case. This can be revealed by considering a modified version in which subsequent consent is withheld. Imagine that, upon regaining consciousness, I object to your surgery on the grounds that my religious commitments preclude medical interventions in my body. It seems clear that my withholding of subsequent consent does nothing to affect the permissibility of the surgery—after all, at the time of the surgery, it was reasonable for you to assume that I had an interest in continuing to exist and would therefore have offered my consent. However, if the denial of subsequent consent does nothing to render the surgery impermissible, then it is not clear what the obtainment of subsequent consent does to render it permissible. Indeed, it seems like what is really doing the work in this type of case is the reasonable expectation that I *would have* consented to the surgery at the moment it was performed, though this appeal to hypothetical consent is not applicable in procreative cases.

One might object that it is not the actual obtainment of subsequent consent that justifies presumptively right-violating conduct, but rather the reasonable expectation that subsequent consent will be offered. Thus, it does not matter that a person actually withholds their consent after the fact so long as it was reasonable to expect that they would provide it. This interpretation of subsequent consent survives the previous objection, but it is also implausible as a general principle, for it would seem to permit right-violating conduct that had the effect of increasing the probability of subsequent consent. For example, suppose that you and your associates find me unconscious one day and perform a sophisticated brainwashing technique that wins my allegiance to your cult. Suppose further that this was the only time the technique could be performed, such that it was not possible to secure my consent beforehand. If the reasonable expectation of subsequent consent is sufficient to render a right-violating act permissible, then it seems that you have acted permissibly, for given the effect of the brainwashing technique, it was reasonable for you to expect that I would later consent to your actions. This, however, is a bizarre implication. It is clearly impermissible for you to brainwash me, and at least one explanation for this is supplied by the inapplicable principle of hypothetical consent: given my interests at the time I became unconscious, it was not reasonable to expect that I would have consented to being brainwashed.

Of course, one might object further that the consent offered in this example would not be morally binding given that it was offered under the judgment-corrupting influence of the brainwashing technique. This line of objection might seem promising, though it does not actually help proponents of the waiving of rights objection, for as many philosophers have noted, human existence contains its own judgment-corrupting influences that predispose people toward existence and make it more likely that they will subsequently consent to having been born into sub-threshold conditions. For example, many philosophers have noted the deceptive role that various psychological phenomena play in our subjective quality of life assessments, including positivity bias, status quo bias, and preference adaptation.[40] One potential effect of these phenomena is to make us subjectively satisfied with unsatisfactory states of affairs, leading us to question whether a child's subsequent consent can be sufficiently informed to justify the violation of their right. However, in addition to these general phenomena, there is also a more specific judgment-corrupting influence at work in the subsequent waiver scenario, which is the false sense of deprivation involved in comparing one's current existence with a counterfactual scenario in which one never existed. When we ask a presently existing person whether they would prefer existing in their disadvantaged condition to having never existed at all, we are asking them a loaded question, for the experience of existence makes it very difficult to imagine non-existence in anything but a negative light. Indeed, by inviting a person to consider a counterfactual scenario in which they never existed, we are implicitly directing their attention toward the various goods in their life they would have "missed out" on, making their counterfactual non-existence seem inherently deprived by comparison. However, in a subsequent waiver scenario, this is problematic for two reasons. First, unlike death, preconception non-existence is *not* in fact a deprived condition, as there are no goods in that condition to be deprived of, and no person to be deprived of them.[41] The sense that it *is* a deprived condition is merely the product of a faulty *ex post* comparison that considers information that is irrelevant to the evaluation of the alternative scenario: had the child never existed, he would not have experienced the goods of existence and would therefore not have been deprived by their absence. Second, to the extent that it is foreseeable that a child *will* view their counterfactual non-existence as a deprived condition, the case of bringing a child into sub-threshold conditions exhibits the same problematic features as the brainwashing case presented above: it is a rights-violation that has the effect of increasing the probability of subsequent consent by means of a judgment-corrupting influence. Thus, the consent that is subsequently offered cannot serve a justificatory role in the violation of that right, as it is again tainted in a way that renders it morally inert.

One might wonder whether this view implies that other proposed standards for valid life-affirmation—such as David DeGrazia's principle of "undeluded gladness"—are difficult or impossible to come by.[42] I am unsure about this strong view, though given the various factors that bias one toward affirming existence even under adverse conditions, I do not think that we can point to a child's subsequent gladness to be alive to justify the violation of his rights. The point is not only that the child might be deluded in a way that invalidates his subsequent consent but also that he is taken advantage of in an unfair way. His parents are able to shirk their duties of procreative care with respect to him in the reasonable expectation he will later absolve them of their wrongdoing, as his experience of existence will predispose him to favor it over a counterfactual scenario in which he never exists. The fact that the deck is stacked in favor of their exoneration should lead us to seriously question the moral relevance of any subsequent consent that is offered.[43]

2.4.4 The Lack of Legitimate Complaint Objection

A related but even deeper worry is that even if a child did not waive their right to a decent start and were to protest its violation, they would nevertheless lack a legitimate complaint against their progenitors insofar as the only alternative to having their right violated was non-existence. This has often been the verdict in so-called "wrongful life" torts, in which a child (or their representative) claims damages against their parents or others for bringing them into a miserable, disadvantaged, or unwanted life. Courts have typically rejected such claims on the grounds that they depend on an impossible comparison between the value of non-existence relative to existence, which does not allow for the establishment of injury on the part of the child. As one judge wrote in response to a case in which a doctor was sued for permitting the birth of a severely disabled child,

> sympathy for a handicapped child and his parents should not lead us to ignore the notions of responsibility, causation, and damage that underlie the entire philosophy of our system of justice. It would be unwise—and, what is more, unjust—to permit the plaintiff to recover damages from persons who caused him no injury.[44]

Now in one sense, this common objection to wrongful life torts simply begs the question against rights-based solutions, as it relies on the contentious "no worse off" argument that such solutions reject. Conceiving of injury as the comparative worsening of one's condition, it deduces that the child has not been injured in wrongful life cases because she cannot be shown to have been made worse off than she otherwise would have been (as such a judgment would entail an allegedly impossible comparison between the value of a defective life and no life at all). But this is the very premise that rights-based solutions seek to discredit. Even if a child cannot be made comparatively worse off by a procreative act, she may nevertheless be wronged if she is born into conditions in which her basic welfare interests cannot be satisfied. Viewed in this way, such a child *does* appear to have a legitimate moral complaint against those responsible for her birth: that her right to a decent start has been violated, or that she has been born into conditions in which that right cannot be fulfilled.

In another sense, however, this common objection to wrongful life torts might indicate a much more problematic notion for rights-based solutions, namely, that there are cases in which a person lacks a legitimate complaint for having their rights violated. As an example of a case that might fit this description, consider another surgery case, based on one originally introduced by Parfit:

> *Accident*: Suppose that you are unconscious after a car accident. Your arm, currently lodged under one of the vehicles, must be amputated imminently in order to save your life. Fortunately, a surgeon happens to be on scene to perform the operation. Knowing what he must do, he successfully amputates your arm, thereby saving your life.[45]

The *Accident* case appears to be one in which the violation of a person's rights does not affect the permissibility of the act in question. It is true that, by amputating your arm without your consent, the surgeon violates a number of your rights, including your right to bodily integrity. However, given the available alternatives, the violation of that right does not provide a strong reason against amputating your arm. If the surgeon does *not* amputate your arm, your right to bodily integrity will be respected, but only at the much higher cost of your life. In this sense, even though the surgeon has violated your rights, you appear to lack a legitimate complaint against him insofar as you would have died otherwise.

An analogy with the *Accident* case appears to undermine the significance of rights violations in procreative contexts. Just as your rights were violated by the surgeon amputating your arm without your consent, so too is a child's right to a decent start violated when he is born into conditions in which his basic welfare interests cannot be satisfied; and just as you lack a legitimate complaint against the surgeon insofar as your life depended on him violating your rights, so too does a child lack a legitimate complaint against his parents insofar as *his* life depended on them violating *his* rights. If sound, this analogy serves as a compelling *reductio ad absurdum*, for if we want to claim that a child does in fact have a legitimate complaint against his parents, we seem forced to take on board the more counterintuitive claim that you also have a legitimate complaint against the surgeon and that his act was therefore impermissible.

However, there are at least two important differences between the case of procreation and the *Accident* case that preclude such an analogy, and which by extension undermine the lack of legitimate complaint objection as a whole. The first difference involves the consequences of not acting, or of refraining from engaging in the right-violating conduct. By not acting in the *Accident* case, the surgeon allows you to die, which results in an even worse outcome from the perspective of your own interests, and which may also violate a general duty of easy rescue. However, by not acting in the case of procreation, there is no comparable loss: the parents would not be allowing a worse outcome to occur—a possible child is not made worse off by never existing—and assuming that there is no duty to procreate, they would not be violating any duties owed to particular people. Thus, while there are strong reasons in favor of acting in the *Accident* case—in light of the consequences of *not* acting—there are only reasons *against* acting in the case of procreation, suggesting that the violation of a child's right to a decent start is not analogous to the violation of your right to bodily integrity. This disanalogy is significant in the present context because it provides the surgeon with a justification for his actions that parents lack: if he does not violate your rights, then things will go worse for you.

However, a second and arguably more important difference between the two cases involves the causal relationship between the right violator(s) and the circumstances of the rights violation. In the *Accident* case, you were by chance put in a position where the only available options were to have your arm amputated or to die, a circumstance that apparently leaves you with no legitimate complaint against the surgeon for violating your rights. However, in the case of procreation, the parents are responsible for *creating* a comparable circumstance. By conceiving amid conditions in which their child's right to a decent start could not be satisfied, they electively put him in a position where he lacks a legitimate complaint against them, insofar as the only alternative to having his right violated was to never exist. A truly analogous *Accident* case would then be one in which the surgeon was responsible for putting you in a position where the only available options were to have your arm amputated or to die—suppose he was in the other vehicle and deliberately crashed into you to perform a surgery. However, in this revised *Accident* case, it would seem odd to attach significance to the fact that you lack a legitimate complaint against the surgeon, for he is responsible for putting you in that compromised position: had he acted with due consideration for your interests, you would not have been in a circumstance where the best available option was to have your arm amputated. In a similar vein, it seems misguided to attach moral significance to a child's apparent lack of complaint against his parents. Had his parents acted with due consideration for his interests—or, more accurately, the interests of whatever child they might have created—he would not have been put in a position where having his right violated was the best available option for him. The fact that he would not have existed otherwise should not get his parents off the moral hook, for they still appear to have done wrong—not only for creating a child whose rights could not be fulfilled but also for putting him in a position where the cost of protesting that injustice is denouncing his own existence.

Of course, when reflecting on this response to the lack of legitimate complaint objection, one might wonder whether it is always appropriate to think of a child's parents as electively putting him in a position where having his rights violated was the best available option for him. For instance, in cases where a child's basic welfare interests cannot be satisfied because of a serious congenital disorder, it might seem more appropriate to say that the disorder, rather than the parents, is responsible for putting the child in a disadvantaged condition. But if this is the case, there seems to be a relevant disanalogy between the case of procreation and the revised *Accident* case presented above, for while the surgeon is responsible for creating the limited option set faced by the accident victim, a set of procreators cannot always be held responsible for the limited option set faced by their progeny.

Does this disanalogy undermine the argument at hand? Not necessarily, for even if procreators are not responsible for the limited option set faced by their progeny, they are still responsible for creating a child who will foreseeably face a limited option set, and this is sufficient to establish the moral wrong. To illustrate the plausibility of this reasoning, consider the following pair of cases. Imagine that I promise to help you paint your house, and then renege on that promise because I am too hungover to perform physical labor. In this case, I have wronged you by failing to fulfill my promise *and* I am responsible for my inability to fulfill that promise—I should not have had so much to drink the night before I agreed to paint your house. Now imagine an alternative case in which I promise to help you paint your house knowing that my heart condition precludes me from performing physical labor. In this case, I am not responsible for my inability to fulfill my promise—this is due to my heart condition—though I have still wronged you in a virtually identical sense by making a promise I knew I would not be able to keep. By parity of reasoning, even if procreators are not directly responsible for the non-fulfillment of their child's rights, they still wrong their child by creating her amid conditions in which they know her rights will go unfulfilled. That this outcome is foreseeable at the time of the child's conception is sufficient to render their act of procreation wrongful.

2.4.5 *The Unfairness Objection*

A fifth and final objection focuses not on the intelligibility of attributing sufficientarian rights to future children, but rather on the potentially unattractive normative implications of doing so.[46] As David Wasserman writes,

> the claim that children are entitled to the satisfaction of such a higher standard [than a life worth living], even if unavoidable impairment precludes it, seems ad hoc, with disturbing implications. It would condemn parents of limited means and prospects, or in hostile or unstable social environments, from having any children, even children expected to have lives worth living despite their privations and hardships. The harshness of this judgment remains even if we excuse parents for the rights violation if they cannot have a child who exceeds the threshold.[47]

Wasserman's worry can be understood as one about unfairness to potential procreators, particularly those who are unable to secure minimally decent conditions for their progeny due to circumstances beyond their control. Does this type of worry provide grounds for rejecting the proposed rights-based solution?

As a preliminary response, it is worth noting that Wasserman does not deny the claim that giving due consideration to children's interests implies that some persons are unable to procreate permissibly—on the contrary, he concedes that persons who are unable to provide their children with "lives worth living" are morally prohibited from procreating, even when that inability is due to circumstances beyond their control.[48] In this sense, Wasserman's objection has more to do with *where* a rights threshold is set, rather than the very idea of a rights threshold itself. Indeed, the crux of his objection seems to be that higher thresholds fail to strike an appropriate balance between the interests of procreators and the interests of progeny and, in so doing, exclude too many people from permissibly having children.

One possible response to this objection is to exempt procreators from the requirement to secure minimally decent conditions for their children when their inability to meet that requirement is due to circumstances beyond their control. This is the approach taken by DeGrazia, for example, who claims that disadvantaged prospective parents are morally "off the hook in those circumstances in which their inability to meet a basic need [for their children] is a consequence of external circumstances and not their own character deficits or pathology."[49] However, it is not clear that Wasserman's objection necessitates this type of conciliatory response. It is certainly regrettable that some persons will, through no fault of their own, be unable to permissibly pursue their procreative projects, and depending on how much value we attach to procreation, this might give us an additional reason to address the various structural injustices that put persons at such a disadvantage. However, provided that we have good independent grounds for attributing to children the rights in question, it is unclear why this provides a reason to reject a rights-based solution to the non-identity problem. Consider, by way of analogy, how this reasoning would reduce to absurdity if applied in a structurally similar context. It is generally accepted that children have a right to a minimally decent upbringing and that parents who are either unable or unwilling to provide such an upbringing are morally ineligible to parent. In this case, attaching a certain amount of weight to children's interests has the effect of disqualifying certain individuals from permissibly parenting, even when their inability to satisfy a child's interests is due to circumstances for which they are not personally responsible, e.g., a lack of parenting competence that is rooted in a traumatic upbringing of their own. However, the bare fact that a particular conception of children's rights disqualifies some persons from permissibly parenting does not necessarily count as a consideration against that conception. While we might have an obligation to offer assistance to such people, or to take certain measures to improve their capacities or resources, it is an odd response to increase their eligibility by weakening our conception of children's rights.

By parity of reasoning, the bare fact that a particular conception of children's rights disqualifies some persons from permissibly procreating does not necessarily count as a consideration against that conception. While it plausibly generates duties of assistance toward such persons, and diminishes the extent to which they are to blame for bringing children into sub-threshold conditions, it does not necessarily impact the separate issue of whether those children have been wronged by being brought into existence.

2.5 The Decent Start Principle

I hope to have demonstrated that the non-identity problem not vindicate, nor insulate from moral criticism, procreative acts that result in children whose basic welfare interests cannot be satisfied for the duration of their childhoods: the fact that such children would not have existed otherwise does not affect their status as rights-holders, nor does it affect the impermissibility of the rights violation, and arguments to the contrary have so far been unpersuasive. However, while an appeal to rights allows us to account for the moral wrong in (some) non-identity cases, the fact remains that preventing the moral wrong requires preventing the very existence of the child who would have been wronged. As an action-guiding principle, this implies the following rights-based directive for prospective procreators:

> *Decent Start Principle*: of all the possible children a set of procreators might create, they may only permissibly create those children whose rights to a decent start will be fulfilled.

The Decent Start Principle specifies the role that prospective children's interests play in limiting the moral permissibility of procreation. It is what we might describe as a wide-child-affecting directive that is based on a narrow-child-affecting account of wrongdoing: it instructs us to create children whose rights to a decent start will be respected rather than numerically distinct children whose rights to a decent start will be violated (as the latter would be wronged through their creation). I would like to conclude my discussion in this chapter by briefly explaining the role of the Decent Start Principle and its relationship to other principles in procreative ethics, particularly impersonal selection principles and principles based on parental role morality.

First, it should be noted that the Decent Start Principle is first and foremost intended to govern what Parfit describes as the different-number choice of whether to have a child.[50] Because it imposes a set of restrictions on the kinds of conditions under which it is permissible to do so, we might refer to it as a *principle of restriction*. Principles of restriction can be contrasted with *principles of selection*, which are meant to govern same-number choices between different possible procreative options. For example, Julian Savulescu's well-known Principle of Procreative Beneficence—which instructs us to select the child out of the possible children we could create who is expected to have the best chance at the best possible life—is a principle of selection, as its role is to tell us which child to create out of the multiple possible children that we could create. It does not on its own impose restrictions on procreation or forbid us from procreating if the available options are sufficiently bad.[51] The Decent Start Principle can be supplemented with a principle of selection, though as a principle of restriction, it always takes priority over selection principles. For example, one might consistently endorse the Decent Start Principle and the Principle of Procreative Beneficence, holding that one has a moral obligation to create the child who has the best chance of the best possible life out of the possible children whose rights to a decent start will be fulfilled. In this case, the Decent Start Principle serves the function of specifying the eligible procreative options to which the Principle of Procreative Beneficence applies, though if there are no available options that satisfy the conditions of the Decent Start Principle, then the Principle of Procreative Beneficence will not come into effect.

Second, it should also be noted that the Decent Start Principle is informed by an account of wrongdoing in non-identity cases that is distinctively *outcome-based*, or focused on the effects of a procreative act on the resultant child. Outcome-based accounts can be contrasted with *input-based* accounts, or accounts of wrongdoing that are focused on the attitudes or intentions with which a procreator has approached a particular procreative decision. For example, Wasserman defends the input-based view that procreators act wrongly when they display attitudes, or act on the basis of reasons, that express a lack of concern or respect for the child who will be brought into existence.[52] Input- and outcome-based accounts often go together—for instance, we arguably express a lack of concern for a child's welfare when we create him knowing that his right to a decent start will be violated—though there are cases in which they come apart. For example, if I procreate negligently without giving due consideration to my future child's welfare, I act wrongly on an input-based account even if my child goes on to live a decent life according to an outcome-based account.

The different roles played by each of these principles suggest that they are not necessarily in competition with one another in a broader account of the morality of procreation. In the case of the *14-Year-Old Girl*, for example, it is not implausible to suggest that there are at least three different objections to the girl's decision: (1) that she has acted on the basis of objectionably selfish reasons; (2) that the child's right to a decent start has been violated; and (3) that she has failed to produce a better off child in his place, which would have contributed to a state of affairs with greater total utility. Recognizing the compatibility of these principles not only serves to support the tenability of the proposed rights-based approach, in the sense that accepting that approach does not require that we reject others, but also contributes toward a more well-rounded understanding of the morality of procreation.

Notes

1 Derek Parfit. *Reasons and Persons* (Oxford: Clarendon Press, 1984), chapter 16. For similar expressions of the same phenomenon, see also Gregory S. Kavka. "The Paradox of Future Individuals," *Philosophy and Public Affairs*, 11(2) (1982), pp. 93–112; and Thomas Schwartz. "Obligations to Posterity" in Brian Barry and R.I. Sikora (eds.) *Obligations to Future Generations* (Philadelphia: Temple University Press, 1978).

2 Dan W. Brock. "Genetic Harms and the Non-Identity Problem: The Case of Wrongful Handicaps," *Bioethics*, 9(3/4) (1995), pp. 269–275, 271.

3 Parfit, *Reasons and Persons*, 351.

4 Parfit, *Reasons and Persons*, 358. Philosophers like Melinda Roberts have pointed out that the *14-Year-Old Girl* is not in fact a pure non-identity case, for the choice the girl faces is not a '2-alternative' choice between having a disadvantaged child or having no child at all, but is rather a '3-alternative' choice between having a disadvantaged child, having no child at all, or having a child but not exposing him to disadvantage (e.g., by adopting him out to more competent parents who will provide him with a better start in life).

Thus, in order for the non-identity problem to genuinely arise in this type of case, we must assume that the child will not in fact have access to better parental care, such that his choice is really between a disadvantaged existence or no existence at all. Notice, however, that we must make similar assumptions in what are commonly regarded as "pure" non-identity cases, such as those involving identity-linked congenital impairments: just as we must assume that the child in 14-Year-Old Girl will not have access to better parental care, we must also assume that there will be no cure or effective treatment for the identity-linked impairment that removes its disadvantage. For Roberts' discussion of the 2-alternative/3-alternative distinction, see Melinda Roberts. *Child versus Childmaker: Future Persons and Present Duties in Ethics and the Law* (Oxford: Rowman and Littlefield, 1998), chapter 3, esp. 94–95.

5 David Boonin. *The Non-Identity Problem and the Ethics of Future People* (Oxford: Oxford University Press, 2014), pp. 127–157; David Heyd. *Genethics: Moral Issues in the Creation of People* (Berkeley: University of California Press, 1992); John A. Robertson. *Children of Choice: Freedom and the New Reproductive Technologies* (Princeton: Princeton University Press, 1994), esp. chapter 4; and Schwartz, "Obligations to Posterity."

6 Heyd, *Genethics*, 99.

7 Parfit, *Reasons and Persons*, chapter 16; Brock, "Genetic Harms and the Non-Identity Problem"; Joel Feinberg. "Wrongful Life and the Counterfactual Element in Harming" in *Freedom and Fulfillment* (Princeton: Princeton University Press, 1992).

8 Feinberg, "Wrongful Life and the Counterfactual Element in Harming," 27.

9 Elizabeth Harman. "Can We Harm and Benefit in Creating?" *Philosophical Perspectives*, 18 (2004), pp. 89–113; and Seana Valentine Shiffrin. "Wrongful Life, Procreative Responsibility, and the Significance of Harm," *Legal Theory*, 5 (1999), pp. 117–148. For a more general discussion and defense of non-comparative harm, see Shiffrin's "Harm and its Moral Significance," *Legal Theory*, 18 (2012), pp. 1–42.

10 Harman, "Can We Harm and Benefit in Creating?" 93.

11 Kavka, 1982; Rahul Kumar. "Who Can Be Wronged?" *Philosophy and Public Affairs*, 31(2) (2003), pp. 99–118; J. David Velleman. "Persons in Prospect," *Philosophy and Public Affairs*, 36(3) (2008), 221–288; and David Wasserman. "The Non-Identity Problem, Disability, and the Role Morality of Prospective Parents," *Ethics*, 116(1) (2005), pp. 132–152.

12 David Archard. "Wrongful Life," *Philosophy*, 79(309) (2004), pp. 403–420; Robert Elliot. "The Rights of Future People," *Journal of Applied Philosophy*, 6(2), pp. 59–69; "Doran Smolkin. "Toward a Rights-Based Solution to the Non-Identity Problem," *Journal of Social Philosophy*, 30(1) (1999), pp. 194–208; Bonnie Steinbock. "Wrongful Life and Procreative Decisions" in Melinda A. Roberts and David T. Wasserman (eds.) *Harming Future Persons* (Dordrecht: Springer, 2009); and James Woodward. "The Non-Identity Problem," *Ethics*, 96(4) (1986), pp. 804–831.

13 Parfit, *Reasons and Persons*, 374; Brock, "Genetic Harms and the Non-Identity Problem," 271.

14 Woodward, "The Non-Identity Problem," 809–811. See also Kumar, "Who Can Be Wronged?" 99–105 and Smolkin, "Towards a Rights-Based Solution to the Non-Identity Problem," 199–202.

15 Woodward, "The Non-Identity Problem," 811.

16 As I explain in further detail below, whether or not a rights-based approach identifies a moral wrong in a particular non-identity case will depend on the substantive content of a child's rights. Thus, a plausible rights-based approach might only identify a moral wrong in a small subset of cases in which the non-identity problem arises. This does not constitute a flaw of the rights-based approach, but simply reflects its limited role in our overall assessment of non-identity cases. In cases that do not involve rights-violations, we may need to appeal to other types of considerations in order to frame a moral objection, e.g., impersonal values or parental virtue.

17 Steinbock, "Wrongful Life and Procreative Decisions," 163.

18 Archard, "Wrongful Life," 406.

19 Joseph Raz. *The Morality of Freedom*. (Oxford: Oxford University Press, 1986), 166.

20 See, for example, David Boonin, *The Non-Identity Problem and the Ethics of Future People*, 109–119 and David Wasserman. "Pro-Natalism" in David Benatar and David Wasserman. *Debating Procreation: Is it Wrong to Reproduce?* (Oxford: Oxford University Press, 2015), 230–236.

21 Feinberg, "Wrongful Life and the Counterfactual Element in Harming."

22 Archard, "Wrongful Life," 404.

23 Even if we make this assumption, notice that the relevant explanation for the absence of wrongdoing is not supplied by the fact that the child is not worse off than he otherwise would have been, but rather that he is not worse off than he is entitled to be.

24 Parfit, *Reasons and Persons*, 364.

25 Woodward, "The Non-Identity Problem," 815. For a similar view, see also Michael Tooley, "Value, Obligation, and the Asymmetry Question," *Bioethics*, 12(2), pp. 111–124, esp. at 119–122.

26 David Heyd. "The Intractability of the Non-Identity Problem," in Roberts and Wasserman, *Harming Future Persons*, 13.

27 Michael Bayles. "Harm to the Unconceived," *Philosophy and Public Affairs*, 5(3) (1976), pp. 292–304, 300. On this point, see also Nils Holtug. "On the Value of Coming into Existence," *Journal of Ethics*, 5(4) (2001), pp. 361–384, 367.

28 See Wilfred Beckerman and Joanna Pasek. *Justice, Posterity, and the Environment: Environmental Ethics for a New Millennium* (Oxford: Oxford University Press, 2001), esp. chapter 2; Richard De George. "The Environment, Rights, and Future Generations" in Kenneth Goodpaster and K.M. Sayre (eds.) *Ethics and Problems of the 21st Century* (Notre Dame: Notre Dame University Press, 1979); and Ruth Macklin. "Can Future Generations Correctly Be Said to Have Rights?" in Ernest Partridge (ed.) *Responsibilities to Future Generations* (Buffalo: Prometheus Books, 1981).

29 Macklin, "Can Future Generations Correctly Be Said to Have Rights?" 151–152.

30 Joel Feinberg. *Harm to Others, Vol 1 of The Moral Limits of the Criminal Law* (Oxford: Oxford University Press, 1984), 97.

31 Parfit, *Reasons and Persons*, 364.

32 See, for example, David Benatar. *Better Never to Have Been: The Harm of Coming into Existence* (Oxford: Oxford University Press, 2006), 22–28; Velleman, "Persons in Prospect," 273–274; Steinbock, "Wrongful Life and Procreative Decisions," 165–166.

33 Benatar, *Better Never to Have Been*, 23.

34 Velleman, "Persons in Prospect," 274.

35 For this type of argument, see Tim Bayne. "In Defense of Genethical Parity," in David Archard and David Benatar (eds.) *Procreation and Parenthood: The Ethics of Bearing and Rearing Children* (Oxford: Oxford University Press, 2010).

36 Boonin, *The Non-Identity Problem and the Ethics of Future People*, 120.

37 Boonin, *The Non-Identity Problem and the Ethics of Future People*, 119–123 and 257–268.

38 For similar arguments, see Jeffrey Reiman. "Being Fair to Future People: The Non-Identity Problem in the Original Position," *Philosophy and Public Affairs*, 35(1) (2007), pp. 69–92, esp. at 88, and Paul Hurley and Rivka Weinberg. "Whose Problem is Non-Identity?" *Journal of Moral Philosophy*, 12(6) (2015), pp. 699–730, esp. at 720.

39 Boonin anticipates this type of response, though denies that it succeeds across the board, for unlike the person who holds a gun to your head, procreators cannot always be held responsible for the limited option set faced by their progeny. I respond to this distinct worry below in connection with the lack of legitimate complaint objection. For Boonin's articulation of it, see Boonin, *The Non-Identity Problem and the Ethics of Future People*, 259–262.

40 For a useful overview of these phenomena and their relevance for procreative decision-making, see Jason Marsh. "Quality of Life Assessments, Cognitive Reliability, and Procreative Responsibility," *Philosophy and Phenomenological Research*, 89(2) (2014), pp. 436–466.

41 Cynthia B. Cohen. "The Morality of Knowingly Conceiving Children with Serious Conditions: An Expanded 'Wrongful Life' Standard," in Nick Fotion and Jan C. Heller (eds.) *Contingent Future Persons: On the Ethics of Deciding Who Will Live, or Not, in the Future* (Dordrecht: Springer, 1997), 33–35.

42 David DeGrazia. *Creation Ethics: Reproduction, Genetics, and Quality of Life* (Oxford: Oxford University Press, 2012), 158.

43 For a related though distinct view about a child's inevitable self-attachment to his own regrettable existence, see Velleman, "Persons in Prospect," 272.

44 Excerpt from New Jersey Supreme Court Justice Schreiber's dissenting opinion in *Procanik v. Cillo* (1984), as quoted in Bonnie Steinbock. "The Logical Case for Wrongful Life," *Hastings Center Report*, 16(2) (1986), pp. 15–20, 18.

45 Derek Parfit. "Comments," *Ethics*, 96(4) (1986), 832–872, 855.

46 See Wasserman, "The Non-Identity Problem, Disability, and the Role Morality of Parents," 137–138, and DeGrazia, *Creation Ethics*, 167–169.

47 Wasserman, "The Non-Identity Problem, Disability, and the Role Morality of Parents," 138.

48 Wasserman, "The Non-Identity Problem, Disability, and the Role Morality of Parents," 135.

49 DeGrazia, *Creation Ethics*, 168.

50 For Parfit's distinction between same-person, same-number, and different-number choices, see *Reasons and Persons*, 355–356.

51 According to this principle, "couples (or single reproducers) should select the child, out of the possible children who they could have, who is expected to have the best life, or at least as good a life as the others, based on the relevant, available information." See Julian Savulescu. "The Principle of Procreative Beneficence: Why We Should Select the Best Children," *Bioethics*, 15(5/6) (2001), pp. 413–426, 415.

52 See Wasserman, "The Non-Identity Problem, Disability, and the Role Morality of Parents," and "Pro-Natalism." Similar input-based accounts have been defended in Kavka, "The Paradox of Future Individuals," Kumar, "Who Can Be Wronged?" and Velleman, "Persons in Prospect."

3 Procreation and Parental Rights

3.1 Introduction

Once a child is born, they require and are morally entitled to adequate parental care for the duration of their childhood. Who, then, should act as the child's parents? In the present and following chapter, I explore the moral relationship between procreation and parenthood by examining the following two questions:

1 Does causing a child to exist through procreation generate a special moral *right* to parent that child?
2 Does causing a child to exist through procreation generate a special moral *obligation* to parent that child?

In the existing literature surrounding the morality of parenthood, many theorists have treated these questions as two sides of the same coin, assuming that the kinds of considerations that serve to ground parental rights are the same kinds of considerations that serve to ground parental obligations.[1] On this view, if causing a child to exist through procreation is sufficient to generate a special right to parent that child, then it is also sufficient to generate a special obligation to parent that child as well. However, as I will demonstrate over the course of the next two chapters, this "parental package" view is in fact false.[2] While procreation can play an important role in the generation of both parental rights and obligations, the precise features of procreation that give rise to parental rights are not the same features that give rise to parental obligations. This means that, in some cases of procreation, parental rights and obligations come apart: one can have a presumptive right, but no obligation, to parent a child one has played a role in creating, and one can have an obligation, but no presumptive right, to parent a child one has played a role in creating, or at least to ensure their adequate parental care. As I will go on to demonstrate, clearly distinguishing the sources of parental rights and obligations is indispensable to resolving a

DOI: 10.4324/9781003164104-4

wide range of procreative conflicts, including cases in which there is disagreement between different parties about who is entitled to parent a particular child (e.g., surrogacy disputes), as well as cases in which there is disagreement between different parties about who is responsible for parenting a child (e.g., cases of accidental fatherhood).

I begin in the present chapter by considering whether causing a child to exist through procreation generates a special moral right to act as that child's parent. In law and common-sense morality, it is widely accepted that adults who meet a minimum threshold of parental competency have a presumptive right to parent their biological children. The legitimacy of this right is also assumed in the parental interest justification of the right to procreate defended in Chapter 1, according to which the right to procreate is grounded in a strong interest that adults have in developing intimate fiduciary relationships with children. However, the philosophical justification of this right is far from clear, for while existing theories of parental rights explain why it is legitimate for adults to exercise authority over children in general, they do not on their own explain how adults come to acquire this authority over specific children. What is needed, then, is a supplementary explanation of why individuals who have a right to parent in general can also be thought to have a special right to enter into parental relationships with their biological children.

I explore this issue in the present chapter by critically examining one explanation that has become prominent in the existing literature surrounding the morality of procreation, and that is closely aligned with the account of the right to procreate developed in Chapter 1: that the right to parent one's biological child is ultimately grounded in an intimate relationship that develops between babies and their birth parents during the process of gestation.[3] The chapter proceeds in five parts. After making some brief terminological clarifications in Section 3.2, I begin in Section 3.3 by outlining the two dominant theories of the right to parent and showing how neither has the resources to explain why adults who have a right to parent in general also have a special right to parent their biological children. I then turn in Sections 3.4 and 3.5 to outline and critique the *relationship-based gestational account* of this right, arguing that it faces at least three serious problems that ultimately discredit it as a tenable account: one relating to its *explanatory* function, or its ability to explain why adults who have a right to parent in general also have a special right to parent their biological children; one relating to its *adjudicatory* function, or its ability to resolve cases in which there are multiple conflicting claims over who is entitled to parent a newborn child; and one relating to its *normative* implications in a series of important test cases. Finally, in Section 3.6, I explore how an alternative autonomy-based approach toward the right to parent one's biological children can avoid these problems. According to the *parental project view*, the right to parent one's biological children is not rooted in any specific feature of the biological relationship but rather in a more general right to initiate and continue one's justice-respecting projects. This right often protects the rights of gestators to parent the children they have gestated, but not necessarily, and not for reasons that make any essential reference to biological connectedness. If sound, this alternative account has important theoretical and practical implications, suggestions among other things that the dominant approach toward surrogacy dispute resolution is currently misguided.

3.2 Disambiguating Parental Rights and Obligations

Before proceeding to the argument, it is first necessary to clarify some of the terminology that I will be using over the course of the next two chapters. Unless otherwise noted, I use the term "parent" to refer to a moral role involving special role-based rights and obligations, rather than to a biological relation between procreators and their progeny. On this interpretation, a parent is someone who has special child-rearing rights and obligations with respect to a particular child. This may be and often is a person who is also biologically related to the child, though it may be a non-biologically related person as well, such as an adoptive parent.

When evaluating the moral role of parents, it is important to disambiguate three distinct senses in which we might speak of parental rights and obligations.[4] First, we might speak of the rights and obligations *of* parents or the various rights and obligations that attach to that particular role. These include, for example, the rights of parents to make certain important decisions on behalf of their children, as well as the obligations of parents to provide their children with certain goods and opportunities. In accordance with one variation of the parental package thesis, the rights and obligations of parents normally do go together as a package, with the acquisition and maintenance of parental rights being conditional on the adequate fulfillment of parental obligations. There are, however, certain exceptions. An abusive parent, for example, may forfeit their right to make important decisions on behalf of their child but retain an obligation to provide their child with material support. Notice, however, that while it is possible for someone to lose parental rights but retain certain parental obligations, there are no parallel cases in which someone can lose parental obligations but retain parental rights. In this sense, the rights of parents normally come packaged with the obligations of parents, even if the latter can sometimes be held independently.

In addition to the rights and obligations *of* parents, we might also speak of the right and obligation *to* parent, or the prior right and obligation that an individual might have to enter into the role of a child's parent. By way of illustration, a man who has agreed to be the godfather of his best friend's child will have both the right and the obligation *to* parent that child in the event of his friend's death, at which point he will then acquire the rights and obligations *of* that child's parent. In this sense, the right and obligation to parent can be understood as preceding the rights and obligations of parents by one step: they are the right and obligation to enter into a role to which certain child-rearing rights and obligations attach. As I indicated above, one of the conclusions that I will be arguing for over the next two chapters is that, contrary to the rights and obligations of parents, the right and obligation to parent do not necessarily come together as a package and are in fact grounded in different types of considerations.

Finally, when we are speaking of the right and obligation to parent, we might distinguish further between the right and obligation to parent *in general*, and the right and obligation to parent a *particular child*. The right to parent in general is, as Sarah Hannan puts it, "the entitlement to vie for, and in some cases, occupy the role of parent,"[5] whereas the right to parent a particular child is the entitlement to occupy that role with respect to a specific child. As Hannan notes, "Establishing that someone has a right to parent [in general] leaves open which specific children they have a claim to parent, and how they come to have claims to parent these children, rather than others."[6]

In the following section, I will explain how this lack of entailment between a right to parent in general and a right to parent a particular child presents a problem for the apparent right of procreators to parent their biological children, though before doing so, it is necessary to make two clarifications. First, because I am concerned with the rights and obligations to parent particular children—and specifically with whether those rights and obligations are generated via procreation—I will not be concerned with elaborating the content of the rights and obligations of parents, apart from what is required for expository purposes. I do not see this as a significant omission, for it is reasonable to think that questions about the content of role-based rights and obligations can be separated from questions about their acquisition. Second, because I do not discuss the rights and obligations of parents in any significant detail, I will from now on reserve the terms "parental rights" and "parental obligations" to refer to the right and obligation to parent, respectively, unless otherwise noted.

3.3 Parental Rights and the Baby Redistribution Problem

The common assumption that adults have a presumptive right to parent their biological children depends on the truth of two premises:

1 adults are generally entitled to stand in authoritative relationships with respect to children; and
2 adults have a special entitlement to enter into those relationships with children to whom they are biologically—that is, either genetically or gestationally—related.

I will demonstrate in this section that while contemporary accounts of parental rights are able to substantiate premise (1), they are ill-equipped to substantiate premise (2). This will highlight the need for a supplementary justification of the right to parent one's biological child, which the relationship-based gestational account promises to provide.

The contemporary literature on the morality of parenthood offers two main accounts of the justification of parental rights: *child-centered* accounts, which justify parental rights solely with reference to the interests of children[7]; and *dual-interest* accounts, which justify parental rights with reference to the interests of children as well as adults.[8] Child-centered accounts begin with the observation that children, as not yet fully formed persons, tend to be poor guardians of their own interests. Left to their own devices, they usually make decisions emotionally and impulsively, and they will often forgo their long-term interests in favor of more immediate desires. In order to develop into healthy adults, then, children require the help of paternalistic caretakers who can make decisions on their behalf and ensure that their long-term interests are being served. Because adults normally require a certain degree of authority over children in order to perform this role effectively, serving as a child's caretaker is also thought to ground a number of limited and conditional rights over that child. Samantha Brennan and Robert Noggle describe this view in the following way:

> Children are immature in a number of ways. Their limited cognitive powers and experience make them prone to mistakes in judging their own interests and how to further them. The healthy mental, physical, and emotional development of children seems to require that someone have the responsibility to nurture and protect the child, and the authority to exercise her own judgment in doing so on a day-to-day basis. Given that someone must do these things, and that children are often too immature to do so, it seems natural to assign parents the right to do so.[9]

One of the attractive features of child-centered accounts is that they fit with common understandings about the conditionality of parental rights. If parental rights are justified indirectly, as it were, as necessary conditions of fulfilling antecedent parental duties, then this explains why parents forfeit their rights when those duties go unfulfilled, e.g., through neglect or abuse. However, while child-centered accounts are well equipped to explain the conditionality of parental rights, their sole focus on children's interests also makes them vulnerable to a well-known objection, which I will refer to here as the *baby redistribution problem*. As Harry Brighouse and Adam Swift put it,

> [I]f all that matters is ensuring that children's interests are met as well as possible, then children should be distributed to those people judged

> most likely to raise them best. If parents' interests play no justificatory role [in an account of parental rights], what would there be to impugn a well-intentioned and efficient government agency that distributed the children, who under a laissez-faire system would be reasonably well-raised, to adults who would be better parents, thus leaving some adequately good parents childless?[10]

The baby redistribution problem suggests that while child-centered accounts provide a coherent justification of the rights *of* parents, or the rights of caretakers to exercise discretion over how they discharge their parental duties, they do not provide a justification of the right *to* parent, or the right of adults to enter into authoritative relationships with children in the first place. And because they do not appeal at all to adults' interests in grounding the right to parent, child-centered accounts are consistent with redistributing children to adults who are best able to serve their interests, including redistributing babies away from their procreators to more competent adoptive parents.

Of course, proponents might respond here that child-centered accounts *do* in fact have the resources to solve the baby redistribution problem. For instance, if the babies are old enough to have formed emotional attachments to their biological parents, then redistributing them to more competent adoptive parents is likely to harm them by imposing separation costs and undermining their interest in continuity of care.[11] Even if the proposal was to redistribute babies at or shortly after birth, there may nevertheless be child-centered reasons against doing so. For example, as I will discuss further in Chapter 4, philosophers like David Velleman and Bernard Prusak have argued that children have an interest in being parented by their biological progenitors that is derived from their interest in having access to their ancestral history or the unique brand of parental love that is made possible in the "given" relationship between procreator and progeny.[12] If these arguments are correct, then they seem to supply us with reasons against baby redistribution.

Notice, however, that even if these arguments are correct (though there are good reasons to be suspicious of them),[13] they do not account for the more fundamental intuition that the baby redistribution problem seeks to pump, which is that redistributing babies away from their custodial parents is unfair *to the parents*, in addition to being potentially unfair to the children. While children have an important interest in having a paternalistic caretaker, adults also have an important interest in the type of paternalistic caretaking that is performed in the context of parent-child relationships.[14] Recall that Brighouse and Swift identify four features of this type of relationship that combine to make it both welfare-generating and non-substitutable, and hence distinctively valuable for adults.[15] First, unlike relationships between adults, the relationship between parents and children is structurally unequal, with children being especially vulnerable to and dependent on the guidance of their parents. Second, the relationship between parents and children is also unlike relationships between adults in that it is always characterized by a type of legitimate paternalism, wherein parents are required to govern children in ways that will promote their immediate and long-term interests. Third, parent-child relationships are also unique in that children lack a determinate conception of their own good, such that parents cannot simply act on the basis of the child's own beliefs about what is good for them, but must also shape (to a certain extent) their beliefs about what is valuable in life. Finally, and most importantly, parent-child relationships are also distinct in the unique brand of intimacy they make possible, wherein the "love that one receives from one's children ... especially in the early years, is spontaneous and unconditional and, in particular, outside the rational control of the child."[16] According to Brighouse and Swift, adults have a strong self-regarding interest in being able to occupy a relationship characterized by these features:

> The [parental] role enables [parents] to exercise and develop capacities the development and exercise of which are, for many (though not, certainly, for all), crucial to their living fully flourishing lives. Through exercising these capacities in the specific context of the intimately loving parent-child relationship, a parent comes to learn more about herself, she comes to develop as a person, and she derives satisfaction that otherwise would be unavailable.[17]

Once we account for the distinctive interest that many adults have in parenting, we can start to see why baby redistribution might be an injustice to parents as well as to children. So long as adults meet a minimal threshold of parental adequacy, their strong interest in having intimate relationships with children weighs in favor of an entitlement to maintain those relationships uninterrupted, even if there are other adults available who might do a better job of serving the child's interests. Importantly, this entitlement is *fundamental* in the sense that it is grounded directly in the parent's, rather than the child's, interests.

The dual-interest account presented by Brighouse and Swift explains why adults have a fundamental right to parent *in general*, and why it would therefore be unjust for the state to redistribute children from what it deems less suitable to what it deems more suitable parents. However, it does not on its own explain how adults acquire rights to parent *particular* children and may therefore be vulnerable to other forms of baby redistribution. Anca Gheaus makes this point by imagining a scenario of "baby scarcity" in which the number of adults who wish to be parents exceeds the number of children who are available to be parented.[18] In this type of scenario, how should we determine who gets to parent which child? The standard response is to claim that, so long as they meet the minimum threshold of parental adequacy, adults are entitled to parent their biological children. Notice, however, that Brighouse and Swift's account does not in fact support such a right. Because their account of the value of parenting does not depend on any biological connection between parent and child, adults who do not have biological children have the same moral interest in parenting as adults who do have biological children, and hence the same fundamental right to parent.[19] Thus, Gheaus imagines two possibilities for redistributing the babies, both of which are consistent with Brighouse and Swift's account:

First, the babies may be redistributed on the basis of parental competence without actually violating anybody's right to parent. If there are more prospective parents than there are children available to be parented, then the situation is simply one in which it is impossible to satisfy everybody's right to parent; and if all adults have an equal right to parent in general, then moral considerations other than the right to parent must be used to determine who gets to parent which child. Gheaus suggests that parental competence may be one such consideration, given that children have a morally relevant interest in having better rather than less good parents.

However, if redistributing babies on the basis of parental competence seems unfair—perhaps because the personality dispositions associated with excellent parenting are in part a matter of brute luck—a second possibility for redistribution is to set up a baby-allocating lottery in which each adult who would make at least an adequate parent is afforded an equal chance of being allocated a baby. In this case, it is possible that babies would in fact be transferred from less good to better parents, though the basis for reallocation would not be grounded in considerations of parental competence, but rather in considerations of fairness.

Both the competence-based and fairness-based redistribution schemes are consistent with Brighouse and Swift's account of parental rights, yet both leave open the possibility that some adults would be denied the opportunity to parent their biological children. This highlights an important limitation of their dual-interest account: while it has the resources to explain why adults have a right to maintain existing parental relationships with the children in their care, it cannot explain why adults are entitled to enter into those relationships with their biological children in the first place.

In the following section, I will outline and critique Gheaus's influential strategy for addressing this limitation, though before doing so, it is worth sidelining a potential objection. One might argue that the specter of baby redistribution is only problematic if there are positive reasons in favor of redistributing babies—otherwise, the mere desire of adults to parent their biological children might provide a sufficient reason against redistribution. This type of argument might seem attractive in the context of laws and social norms that give adults a strong presumptive claim to parent their biological children, though there are at least two problems with it upon reflection. First, even if there were no positive reasons in favor of redistributing babies, it is still problematic that an account of parental rights permits it in theory, for it suggests that one of the most widely accepted and deeply valued parenthood-related rights is rooted only in convention and has no normative basis. However, a second and arguably more significant problem for this type of argument is that there *may* in fact be good reasons for redistributing babies under certain circumstances. For example, Gheaus suggests that in societies with entrenched histories of racism or sexism, redistributing babies randomly at birth might have the effect of improving equality of opportunity by mitigating the influence of race or gender on one's life prospects.[20] Moreover, if we take the opportunity to parent as itself a distribuendum of justice—as philosophers like Brighouse and Swift do—then redistributing babies randomly between all adequate prospective parents might also be a fairer way of allocating parenting opportunities, particularly for those who might be at a disadvantage pursuing parenthood under a *laissez-faire* system, including same-sex couples and single or infertile individuals. This, of course, is not to say that pursuing either of these ends justifies baby redistribution, but only that such a conclusion is tenable in the absence of a fundamental right to parent one's biological baby.

3.4 The Relationship-Based Gestational Account

If the arguments in the previous section are correct, then neither the child-centered nor the dual-interest account of parental rights necessarily supports a fundamental right to parent one's biological baby. A child-centered account might be able to explain why it would be unfair to children to redistribute babies away from their biological parents, though this does not establish a *right* to parent one's biological child—at most, it simply establishes an *obligation* to do so, and only if we accept that children have a morally relevant interest in being parented by their biological progenitors. A dual-interest account can explain why it would be unfair to parents to disrupt *already existing* parent-child relationships, though, to the extent that their interest in those relationships is conceived independently of any biological connection to the child, it cannot on its own explain how adults acquire rights to parent their biological children in the first place. This, it seems, leaves us with one of two options: we can either (1) bite the bullet and accept that there is no fundamental right to parent one's biological baby or (2) explain how such a right can be incorporated into the dual-interest framework.

Given the widespread acceptance of the right in question, option (1) is not very attractive. Even if it were in a child's best interest to be handed over to a set of excellent adoptive parents upon its birth, forcibly redistributing babies away from their adequate biological parents still seems like an injustice to those parents, and this is a difficult intuition for many people to abandon.[21] Moreover, pursuing option (1) would also require a radical revision in our understanding of the right to procreate, for as I argued in Chapter 1, the value of that right is conditional on the ability to parent the children created as a result of its exercise. However, option (2) is not without its difficulties either, for attempts to ground a rights-claim in the fact of biological connectedness have sometimes run the risk of collapsing into the kind of "proprietarian" account that many philosophers now consider indefensible. To take one commonly discussed example, some have argued in accordance with a Lockean theory of acquisition that insofar as individuals acquire rights over the fruits of their self owned labor, so too may procreators acquire rights over the fruits of their procreative labor.[22] This type of argument might be able to explain why individuals have a right to parent their biological children, though it does so at a cost, for many take it to imply that children, like Locke's acorns and apples, are simply things to be owned.

Recognizing this difficulty, Gheaus has sought to provide an alternative account of the right to parent one's biological baby that does not appeal to a proprietarian claim over one's offspring, but rather to the significance of the relationship that is established between procreators and their progeny during the process of gestation. This gestational account has emerged as the most sophisticated account of the right to parent one's biological baby in the existing literature, so it is worth exploring in some detail.

Gheaus's account begins with the observation that pregnancies always impose a variety of burdens (or "costs") on pregnant women and their supporting partners. For instance, throughout the duration of their pregnancies, women endure numerous *physical* burdens, from the aches, pains, nausea, and fatigue that are experienced during gestation to the intense pain and physical trauma that often accompanies childbirth. These burdens are not only intrinsically costly in terms of being physically unpleasant but also instrumentally costly in terms of curtailing women's autonomy, making them increasingly reliant on others for the performance of everyday tasks. Moreover, women also endure a number of *social* and *behavioral* costs as a result of carrying out a pregnancy. They face limitations on the food they can eat, the amount of alcohol they can consume, and the physical activities they can safely engage in, and they must cope with the patronizing and sometimes intrusive behavior of friends, family, co-workers, and strangers. Finally, pregnant women also endure a wide range of *psychological* and *emotional* costs prior to giving birth. They must deal with the constant fear of miscarriage or of something else going wrong with the pregnancy, and they must sometimes confront the daunting decision of whether to continue a pregnancy in the face of significant health risks.[23]

One might think that the mere incursion of these costs is sufficient to ground a desert-based claim to parent the resultant child. Joseph Millum, for example, defends what he calls the *investment theory* of parental rights, according to which "parental rights are generated by the performance of parental work."[24] On the basis of this type of theory, one might reason that because pregnant women and their supporting partners incur significant labor-related costs as a result of bringing a child into existence, they are more deserving than other available adults to act as that child's parents. However, Gheaus does not believe that this type of argument can perform its intended function, for while the incursion of costs may entitle procreators to parent *a* baby, it does not necessarily entitle them to parent their biological baby in particular. Indeed, appealing to costs alone would not preclude a different type of redistributive scheme in which newborn babies were shuffled randomly between all eligible procreative couples.[25]

However, while a strictly desert-based argument ultimately fails, Gheaus believes that the costs of pregnancy are significant for a second reason, namely, because their incursion "facilitates the creation of an intimate relationship between the bearing parents and the future baby."[26] As she explains,

> Because children come into existence through gestation, pregnant women and their supporting partners have to invest a significant amount of resources into having birth children. This is often a conscious, intentional process, akin to other projects in which people engage: it contains much anticipation and planning, thinking and hoping, imagination and projection. Through their bodily connection with the baby and their various psychological investments, expecting parents normally build a relationship with their future baby, which is something highly emotional and already quite developed at birth.[27]

According to Gheaus, the relationship that develops between procreators and their progeny during gestation provides the "missing step" in the justification for a fundamental right to parent one's biological baby. While existing dual-interest theories explain why it is impermissible to disrupt already existing parent-child relationships, they do not on their own explain why adults are entitled to enter into those relationships with their biological children in the first place. Yet if the process that brings children into existence also results in the development of an intimate relationship between the baby and its birth parents, then birth parents do not need a special justification for initiating a relationship with their baby, as such a relationship has already been initiated by the time the baby is born. As Gheaus concludes:

> If, at the moment of birth, adequate parents have already paid significant costs for becoming parents, and in the process have developed an intimate incipient relationship with the baby, they are more entitled than other prospective parents to parent the baby they have borne. The difference between adequate bearing parents and other adequate prospective parents can provide the necessary justification for translating the fundamental right to parent in general into a right that birth parents have to parent their birth baby.[28]

Gheaus's account has a number of attractive features. First, it seems to provide a coherent justification for the right to parent one's biological baby without collapsing into a form of proprietarianism. Procreators are not entitled to parent their biological babies because they have general ownership rights over their offspring, but rather because they tend to develop morally significant relationships with their babies during the process of gestation. Moreover, this justification fits neatly into the dual-interest framework, as it appeals to the same kinds of considerations that dual-interest theorists appeal to in order to justify the right to parent in general. The reason why it would be unjust to redistribute babies away from their adequate biological parents is the same reason that Brighouse and Swift provide as to why it would be unjust to redistribute children away from their adequate custodial parents: it disrupts an already existing relationship that the parents (and perhaps also the children) have a morally relevant interest in continuing. Finally, Gheaus's account might also be considered advantageous insofar as it provides a philosophical justification for existing legal norms. In parental custody disputes involving gestational surrogacy and donated gametes, courts and legislators in Europe and North America have typically assigned legal motherhood to the woman who has gestated the embryo, rather than the woman who donated the ovum or commissioned the surrogacy arrangement.[29] Gheaus's account has the resources to explain why these judgments may be correct.

However, while Gheaus's account presents a number of advantages over existing accounts of the right to parent one's biological baby, it also runs into at least three serious problems: one relating to its *explanatory* function, or its ability to explain why adults who have a right to parent in general also have a special right to parent their biological child; one relating to its *adjudicatory* function, or its ability to resolve cases in which there are multiple conflicting claims over who gets to parent a newborn child; and one relating to its *normative* implications in a series of important test cases. I will review each of these problems in turn before outlining an alternative strategy for addressing them.

3.5 Three Problems

3.5.1 The Explanatory Problem

The first set of difficulties facing Gheaus's account stems from its reliance on the realization of certain qualitative features in a pregnancy, particularly the development of an intimate maternal-fetal bond. This is problematic for at least two reasons. First, it seems to carry the odd implication that in cases where such a bond is *not* established, gestating women do not have fundamental rights to parent their biological children. Imagine, for example, a woman who suffers from antenatal depression throughout her pregnancy, a condition that can be characterized by (among other things) feelings of physical and emotional detachment from one's fetus. This woman may not have developed an intimate bond with her fetus, though given that she has incurred all of the typical costs associated with carrying out a pregnancy (in addition to the ancillary cost of coping with her depression), it would still seem unjust to deny her the opportunity to parent her biological baby. Indeed, the only difference between this woman's pregnancy and a non-depressed woman's pregnancy is a particular attitude felt toward the fetus and one that is itself the product of an unchosen psychological affliction.

Gheaus's response to this difficulty is more or less to bite the bullet: she concedes that such a woman would not have a fundamental right to parent her biological child, though denies that this is a significant problem for her account, as "the existence of an intimate relationship between birth parents and newborns is the rule rather than the exception."[30] However, even if we bite the bullet about atypical pregnancies, there is still a second and more serious problem with the reliance of Gheaus's account on maternal-fetal bonding, which is that it is controversial that such a bond *ever* materializes, at least not in the rights-grounding sense that Gheaus describes. As Lindsey Porter has argued, empirical research from nursing and psychology suggests that the apparent relationship that Gheaus describes between pregnant women and their fetuses is really better characterized as a one-directional attitude.[31] While it is common for pregnant women to self-report feelings of attachment to their fetuses, these feelings have been explained by psychologists as responses to social or cultural cues, rather than as evidence of a two-way maternal-fetal bond: there is in fact no physiological evidence to suggest that fetuses share in the affective attitudes of their gestators, nor is there any evidence to suggest that maternal feelings of attachment have any effect on developing fetuses. Thus, "[g]iven that the woman's feelings aren't affecting the fetus, and the fetus's feelings aren't affecting the woman," Porter concludes that "it is implausible to suppose that the right way to characterize MFA [maternal-fetal attachment] (or 'bonding') is as a relationship. It's not a relationship; it's an attitude."[32]

If Porter is right that maternal-fetal bonding is better characterized as an attitude rather than a relationship, then the relationship-based gestational account is in serious trouble. The key selling point of this account is that it purports to offer a distinctively relationship-based account of the right to parent one's biological baby, one that explains why redistributing babies away from their adequate biological parents is just as wrong—and wrong for the same reasons—as redistributing children away from their adequate custodial parents. However, if the crucial maternal-fetal bond that drives this account is really just a one-directional attitude, then the case of gestation is not really analogous to existing parent-child relationships in which *both* parent and child are interpersonally engaged. Rather, it is more analogous to a case in which a person simply *feels* attached to something that is not capable of reciprocation.

There are two possible ways that Gheaus could respond to this difficulty, though neither is particularly helpful for her account. First, she might claim that a one-way feeling of attachment is in fact sufficient to ground a right to parent one's biological baby. This would defuse the worry that mothers do not develop relationships with their fetuses *per se*, though if merely feeling attached to something is sufficient to ground a right to that thing, then this strategy also invites a slippery slope toward some rather dubious rights-claims.[33] Alternatively, Gheaus might concede that a one-way feeling of attachment is insufficient to ground a right to parent one's biological baby, but insist that there are still good child-centered reasons against separating babies from their gestational mothers (to the extent that such reasons are available).[34] This would provide a solution to the baby redistribution problem, but only on child-centered terms—it would not explain why redistributing babies away from their biological parents is an injustice to the parents and so would not constitute an argument for a fundamental right to parent one's biological baby.

3.5.2 *The Adjudicatory Problem*

The explanatory problem poses a challenge specifically for an account that seeks to ground parental rights in a two-directional maternal-fetal relationship. One might object, however, that this is not the only way to ground parental rights in gestation. For example, Millum's investment theory seems to provide a way of grounding parental rights in gestation that does not involve assuming that a special relationship exists between pregnant women and their gestating fetuses—rather, it only requires that we accept (a) that an agent's stake in an entity is proportional to the amount of appropriate work she has put into that entity and (b) that pregnant women perform the highest proportion of appropriate work in bringing a child into existence.[35] However, there is a second set of problems that faces *any* account of parental rights that assigns special normative significance to gestation, one that emerges from what might be described as a dilemma regarding their inclusivity. Even if gestation has the significance that Gheaus ascribes to it, one glaring problem with appealing to gestation as a basis for the right to parent one's biological baby is that it seems to exclude men from the constituency of persons who are eligible for that right. If the labor and bonding that occurs during gestation is what grounds parental rights, then it seems like only women can have fundamental rights to parent their biological children (as long as only women are capable of gestation). This, in turn, suggests an important limitation of Gheaus's account: while it might have the resources to explain why it would be unjust to redistribute babies away from their adequate biological mothers, it seems ill-equipped to explain why it would be unjust to redistribute babies away from their adequate biological fathers.

One could of course bite the bullet on this issue and simply accept the implication that men do not have fundamental rights to parent their biological children. This is the route taken by Barbara Katz Rothman, for example, who insists that "If men want to have children, they will either have to develop the technology that enables them to become pregnant … or have children through their relationships with women."[36] However, while biting the bullet is a viable option, it is not a very attractive one. Not only does it seem intuitively unfair that only women should have fundamental rights to parent their biological children, but it also violates what some parental rights theorists call the Parity Principle, which holds that "any fact by virtue of which a woman laid claim to be a parent could also be a fact in virtue of which a man with equal merit could claim to be a parent, and vice versa."[37] If we believe that an account of parental rights ought to respect the Parity Principle—perhaps in virtue of the equal interest that men and women have in parenting children, combined with the fact that we are not responsible for our biological childbearing capacities—then it seems like we have a *prima facie* reason to reject an account that grounds parental rights in gestation.

Gheaus acknowledges this worry, though rather than rejecting the Parity Principle, she attempts to accommodate it by showing how men (and non-gestational partners in general) can also share in the labor and bonding that generates parental rights. As she writes,

> Like in the case of paying the costs of pregnancy, pregnant women's supporting partners are capable of being direct participants in the process of

> creating a relationship with the baby during pregnancy. With the help of medical technology, they can see the fetus and hear its heartbeat as early as the bearing mother; during the last stages of pregnancy they can feel the baby, talk to it and be heard by it. Just like the mother, they can experience the fears, hopes, and fantasies triggered by the growing fetus.[38]

On this view, even though men are not capable of gestating babies themselves, they are still capable of participating in many of the relationship-building activities that women typically engage in over the course of their pregnancies, which may be sufficient to generate a relationship-based right to parent the resultant child. This type of approach might allow Gheaus to account for the rights of biological fathers, though it ends up introducing a new set of problems, for if male partners can acquire parental rights in virtue of participating in the various relationship-building activities that occur during a pregnancy, then it seems as though a laundry list of other persons can as well.[39] For example, in addition to their male partners, many pregnant women also receive support from their siblings, parents, relatives, and friends, each of whom participates in the various relationship-building activities that occur during a pregnancy, and sometimes to an even greater extent than the baby's biological father. If we want to argue that men can acquire parental rights by virtue of their proximity to their gestating partner, then we risk opening the door to a host of other potential rights-claimants as well.

Gheaus might simply concede that, along with male partners, siblings, parents, relatives, and friends can also acquire rights to parent children with whom they have sufficiently bonded during gestation (even if they do not always assert those rights). However, this is not an attractive option. Not only does it seem intuitively implausible that so many people could have legitimate claims to parent a particular child,[40] but acknowledging any parental rights-claims on the basis of mere proximity to the gestator also prevents Gheaus's account from serving a crucial *adjudicatory* function in cases where there are multiple conflicting claims over who gets to parent a particular child. To understand why, consider an important type of case in which such a conflict may arise:

> *Gestational Surrogate*: Alice and Bill desperately want to have a biological child, though Alice suffers from a form of infertility that makes her unable to safely gestate an embryo. Eager to fulfill their procreative aspirations, Alice and Bill create an embryo using *in-vitro* fertilization and arrange for a gestational surrogate, Claire, to bring the embryo to term. However, after nine months of pregnancy, Claire cannot bear the thought of relinquishing the child, and now wants to keep the baby against her prior agreement with Alice and Bill.

Initially, it might seem like Gheaus's account has the resources to settle this type of dispute—indeed, if parental rights are grounded in gestation, then Gheaus's account seems to provide a straightforward explanation for why the surrogate, rather than the gamete providers, should be entitled to act as the child's parent. Notice, however, that if Gheaus wants to maintain her initial response to the worry about parental parity, then this resolution to the surrogacy dispute does not necessarily follow. In the same way as male partners can participate in the various relationship-building activities that occur during a pregnancy, so too can commissioning parents in surrogacy contracts: they can accompany the surrogate to doctor's appointments, they can view ultrasound scans of the developing fetus, they can feel its kicks and rumblings inside the surrogate's abdomen, they can talk to and be heard by it, and they can share in the various fears, hopes, and fantasies triggered by the growing fetus. If male partners can acquire parental rights in virtue of their participation in these relationship-building activities, then it is difficult to see how Gheaus's account can consistently exclude commissioning parents.

The inclusivity dilemma facing Gheaus's account can now be stated as follows: either parental rights are acquired via gestation, in which case men do not have rights to parent their biological children; or men can acquire parental rights through their proximity to the gestator, in which case so can a host of other potential rights-claimants that a gestational account is wont to exclude. Both horns of this dilemma are unattractive—the first because it is intuitively under-inclusive and the second because it is intuitively over-inclusive—though in its present formulation, Gheaus's account cannot avoid choosing between them.

3.5.3 The Normative Problem

Confronted with the inclusivity dilemma, a proponent of the relationship-based gestational account might choose to reject the Parity Principle and concede that only women have fundamental rights to parent their biological children. This, it seems, is the lesser of two evils: while rejecting the Parity Principle yields the surprising conclusion that men do not have fundamental rights to parent their biological children, maintaining it yields the even more problematic conclusion that those rights can be claimed by anyone who has sufficiently bonded with a baby over the course of a pregnancy. Not only is this conclusion impracticable in its own right, but it also precludes the relationship-based gestational account from serving a crucial adjudicatory role in precisely the kinds of cases we need it to perform that role, namely, cases in which non-gestational bonders also assert the right to parent a newborn child.

However, while rejecting the Parity Principle preserves the adjudicatory function of Gheaus's account, it also exposes a new set of problems, for the reasoning that is used to justify the assignment of parental rights to the gestator in cases like *Gestational Surrogate* appears to be based on an important oversight in its application of the interest theory of rights. At the most general level, Gheaus's account relies on the interest theory supposition that having a strong interest in a thing is sometimes sufficient to generate a right to that thing.[41] This supposition is often true: there are many cases in which we acquire rights to certain things or states of being in virtue of the contribution those things or states of being make to various aspects of our well-being. However, a common limitation or exception to this principle involves cases in which another person already has a legitimate claim to the thing in question. To take a somewhat trivial example, imagine that I hire you to look after my flower garden while I am recovering from an illness. In the following months, you become surprisingly invested in it: you enjoy the ritual of watering and fertilizing it, you take pleasure in watching it grow, and caring for it has uncovered an innate set of skills you were previously unaware you had, including patience, attention to detail, and a keen sense of aesthetic design. While you may have a strong interest in tending to the garden upon my recovery, it does not follow that you thereby have a right to do so, even if you stand to benefit from it more than I do. Your strong interest in the garden provides me with a reason to let you care for it, though it is not in itself sufficient to hold me under a duty to do so, particularly if discharging that duty requires abdicating my own role in caring for it.

Now, there are obviously many ways in which a flower garden is unlike a fetus, and I will be careful to account for these differences in further detail below. However, at a very general level, it seems like the same kinds of considerations that undermine your claim to care for the garden against my wishes can also be invoked to undermine the gestational surrogate's claim to keep the baby against the wishes of the commissioning couple. Presumably, the reason why your interest in the garden does not justify a right to care for it is because I already have a pre-existing claim to it: assuming that my gardening project was justly initiated and continues to be justice-respecting (e.g., it is carried out on a plot of land to which I have legitimate access, using seeds that have been legitimately acquired, without taking up space that is otherwise needed for essential food production, etc.), it is one that I have a right to continue without interference from other people. By parity of reasoning, if the commissioning couple can be thought to have developed a morally relevant interest in the embryo they have created, then the reason why the gestational surrogate's interest in the baby does not justify a right to keep it is that the commissioning parents already stand in a special rights position with respect to it themselves. It was created out of gametes to which they had legitimate access for the purpose of serving *their* parenting project, so they should have a presumptive right to continue that project without undue interference from others.[42]

Of course, intuitions vary about gestational surrogacy disputes, making them non-decisive test cases for the defensibility of this type of reasoning. So consider another important test case, one that shares relevant features of a gestational surrogacy dispute, but where our intuitions might be more settled:

> *Gamete Thief*: Dana breaks into the freezer of a gamete storage facility and steals an embryo that Ernest and Fran had frozen for a future procreative project. She then has it implanted in her uterus and gestates it for the following nine months. When Ernest and Fran discover what Dana has done, they confront her at the hospital on the day the baby is born and plead that she hand it over to them to parent.

According to Gheaus's gestational account, Dana is in the strongest rights position with respect to the baby in virtue of the fact that she has gestated it for nine months and participated in the various relationship-building activities that give rise to parental rights. However, this seems entirely backwards. Rather than acting in a way that makes her eligible for the acquisition of parental rights, Dana appears to have engaged in an egregious form of wrongdoing: she has gestated an embryo that in some very basic sense was not hers to gestate. Notice, however, that we can only account for this intuition if we accept the claim that Ernest and Fran already stand in a special rights position with respect to the embryo they have created—otherwise, there is no principled basis on which to challenge Dana's parental rights-claim to the baby it grows into.

With a few negligible differences,[43] the case of *Gamete Thief* is structurally similar to the case of *Gestational Surrogate*, such that the reasoning we use to resolve one case ought to be applied *ceteris paribus* to the other. If we think that Claire has a gestationally based right to parent the baby in *Gestational Surrogate*, then we are forced to extend the same reasoning to Dana in *Gamete Thief*. If, however, we think that Dana's parental rights-claim is illegitimate in virtue of her relationship to the embryo, then we ought to think the same of Claire, who differs from Dana only in terms of the point in the gestational process at which she decided to "steal" the embryo. I think we have good theoretical reasons to endorse the latter view and thereby reject the principle that parental rights are always acquired via gestation. This does not imply that gestation is irrelevant to the acquisition of parental rights, though it does suggest that it cannot be the whole story and that we may have to look elsewhere for a principle that better systematizes our judgments about these different cases.

I will propose one such principle in the following section, though before doing so, it is worth considering two potential objections that might be raised against the reasoning I have deployed in this section. When confronted with the case of *Gamete Thief*, a proponent of the relationship-based gestational account might respond in the following way:

> *it is true that by gestating Ernest and Fran's embryo without their consent, Dana has violated their pre-existing claim to it and has thereby engaged in a form of wrongdoing. It does not necessarily follow, however, that Ernest and Fran are now entitled to parent the resultant child. While Dana owes Ernest and Fran compensation for stealing their embryo, she may still be in the strongest rights position with respect to the child in virtue of the relationship that has been established during the process of gestation.*

This objection suggests that there is an explanatory gap between the premise that Ernest and Fran have a pre-existing claim to their embryo and the conclusion that they are entitled to parent the child it grows into. Even if we accept that they have such a claim, it is possible that other moral considerations come into play during the process of gestation that diminish its significance and ultimately weigh in favor of assigning parental rights to the child's gestator.

Is this objection sound? I believe whatever force it has stems from an unstated but implicit assumption that Dana, by virtue of having gestated the child, is better qualified to act as that child's parent. So, to fix ideas for a moment, let us assume that there is in fact no discernible difference in parenting competency between the two conflicting parties, such that the scenario is one in which we need a tiebreaker principle to determine who is entitled to parent the child. When the facts of the case are fixed in this way, it seems clear that the balance of reasons weighs in favor of assigning parental rights to Ernest and Fran. Dana may have formed parental intentions with respect to the baby she gestated and, in so doing, developed a strong interest in parenting it upon birth. However, because this interest was formed illegitimately in relation to an embryo to which she lacked legitimate access, it is not one that should carry any moral weight in the assignment of parental rights. And once we discount Dana's interest from our moral assessment, the case becomes easy to resolve: while we have a weighty reason to assign parental rights to Ernest and Fran in light of the legitimate parental intentions they have formed with respect to their embryo, we have no weighty reason to assign parental rights to Dana.

Of course, a proponent of the relationship-based gestational account might object to this reasoning on the grounds that it relies on a philosopher's abstraction—namely, the stipulation that there is no relevant difference in parenting competency between the gestator and the commissioning couple. Even though Dana's parental interest has been formed illegitimately, and does not itself merit any moral consideration, it might be argued that she still possesses gestation-related attributes that provide child-centered reasons against a transfer of custody. For instance, suppose that Porter is wrong and that babies really do develop something akin to a relationship with their gestator over the course of a pregnancy. If this is the case, then transferring a baby away from its gestator will always involve harmful separation costs for the baby. Thus, even if we discount the gestator's interest in parenting the baby, they will always be in a privileged rights position with respect to the baby simply by virtue of how babies come into the world.[44]

Is this further objection sound? Setting aside the fact that it could only support a child-centered, rather than fundamental, right to parent one's gestated child, I believe that it fails for at least two reasons. First, the reasoning that underlies it appears to reduce to absurdity when applied in structurally analogous cases. Consider, for example, one such case:

> *Baby Snatcher*: George longs to become a father, though he is unable to find a procreative partner and his adoption applications have consistently been rejected on account of his being a single male. Frustrated with the biases in the adoption system, and desperate to fulfill his parental aspirations, George snatches a baby from a hospital nursery who has just been born to a pair of loving parents. When the authorities apprehend George a year later and seek to reunite the baby with its original parents, George protests on the grounds that it would impose unreasonable separation costs on the child, who now enjoys an intimate and caring relationship with George.

If the costs of separation are sufficient to preclude a transfer of custody in the case of *Gamete Thief*, then surely they are sufficient to preclude a transfer of custody in the case of *Baby Snatcher*, where the relationship in question is more fully developed and not subject to the controversy surrounding maternal-fetal bonding. This, however, is a troubling implication, for it suggests that one can legitimately acquire the right to parent a child via clearly illegitimate means, so long as the child in question develops an interest in the relationship being maintained.

However, a second and more important reason why the objection fails is because the separation costs to the baby cannot plausibly be appealed to in order to preclude a transfer of custody. In fact, the claim that they *can* be must rely on one of two untenable premises: either (1) a deeply implausible conception of the harm associated with being separated from one's gestator or (2) a deeply implausible weighting of competing interests. In order for it to be the case that the separation costs to the baby preclude a transfer of custody, the harms must be so severe as to outweigh any parental interest on the part of the original intended parents, or the parental interest must be so weak or insignificant as to be overridden by even non-serious harms. Neither of these claims are defensible. While there may be immediate and short-term harms associated with separating a baby from its gestator, it is implausible to suppose that these harms are so severe as to render the original intended parents inadequate[45]—they are at best a consideration that acts as a tiebreaker between the gestator and the intended parents, giving the former an edge over the latter from the perspective of the child. However, if the harms of separation are not so severe as to factor into the adequacy of the intended parents, then the objection relies on a seriously implausible weighing of interests, one that expresses a similar lack of sensitivity toward the interests of parents as the purely child-centered accounts of parental rights canvassed in Section 3.3: it assumes that the avoidance of short-term, non-serious harms to the baby justifies forever denying Ernest and Fran the opportunity to parent the child they deliberately set out to create.[46] The implausibility of this view is compounded when we consider that Dana is responsible for the fact that the child now faces these harms: had she not stolen the embryo and allowed Fran to gestate it, the resultant child would not be in a position where he faces the costs of separation.

If these arguments are correct, then we ought to reject the principle that parental rights are always acquired via gestation. While a gestator may develop a strong attachment to a baby over the course of a pregnancy—and hence a strong interest in parenting the baby upon its birth—this interest does not necessarily carry any moral weight in cases where other parties already have a legitimate interest in parenting the baby themselves. This suggests that there is a more fundamental principle underlying the right to parent one's biological child, one that is lexically prior to gestation and that works to determine whether gestation has normative significance in individual cases.

3.6 The Parental Project View

What, then, could this principle be? Following a broadly liberal approach toward the acquisition of parental rights, I believe that a promising candidate can be found in what I will call the general autonomy principle or the notion that individuals have a strong presumptive right to non-interference in their justice-respecting projects. That individuals have such a right is a familiar component of liberal theories of justice, particularly those that prioritize individuals' ability to formulate, revise, and pursue their own conceptions of the good, though its relevance for the acquisition of parental rights has yet to be fully explored.[47] A full exploration is perhaps beyond the scope of this chapter, though I hope to demonstrate its promise in explaining why adults who meet a minimum threshold of parental competency normally have a presumptive right to parent their biological children. This explanation depends on a number of contestable assumptions about justice in child-rearing, and so is presented conditionally, though if sound it has the considerable advantage of being rooted in a widely accepted moral principle *and* avoiding the three major problems facing the relationship-based gestational account.

To understand the appeal of this type of account, let us return for a moment to the case of *Gamete Thief*, which was originally presented to show that gestating a child is neither necessary nor sufficient for the acquisition of parental rights. In this case, Ernest and Fran are wronged by the theft of their embryo, and in a way that is not reducible to a violation of their property rights—they are also wronged in a more fundamental sense by having an important project of theirs interrupted. By deliberately creating and storing their embryo with the intention of parenting the child it eventually grows into, Ernest and Fran have embarked on a central life project, one they are permitted to begin, and entitled to continue, without interference by other people or institutions. They are permitted to begin their project because they are not (by stipulation) under a duty *not* to begin it: by creating an embryo out of reproductive gametes to which they have legitimate access, they are not infringing on the justice-based entitlements of any other person. They are entitled to continue their project because (a) its continuation does not (by stipulation)[48] infringe on the justice-based entitlements of other people, and (b) the legitimate parental intentions they have formed with respect to their embryo serve to *create* duties of non-interference in others. They normatively transform the embryo from a group of cells over which they have property rights to a central component of their conception of the good, one whose pursuit others—including Dana—are duty-bound to respect. This principle simultaneously explains why Dana's parental rights claim is invalid: while she may see herself as having embarked on her own project by gestating and giving birth to Ernest and Fran's embryo, it is not a project that meets the criterion of being justice-respecting if we accept the plausible claim that individuals have initial rights to control the use of their reproductive gametes.[49]

A similar line of reasoning can be invoked to support the right to parent one's biological child in more familiar cases. In paradigm cases of procreation, two consenting adults deliberately combine their genetic material to create a child who they plan to jointly parent. From the moment a pregnancy is confirmed, they engage in a series of actions and decisions that reflect their commitment to that project and solidify its place within their broader conception of the good: they purchase baby clothes and accessories; they prepare a nursery in their home; they inform their friends and family of the impending arrival; and they rearrange their personal and professional lives to accommodate their upcoming parental responsibilities. We do not need to assume that a relationship exists between a baby and its birth parents to ascribe significance to this project nor do we need to assume that there is anything special about biological connectedness *per se*. The process of legitimate project-formation has moral significance independently of these factors and serves to set a child's procreative parents apart from other prospective parents: *they*, and not others, have engaged in a process of legitimate project-formation with respect to the child they have created, so *they*, and not others, should have a presumptive right to carry out that project by parenting the child upon birth. One important feature of this account is that biological connectedness is only derivatively significant for the acquisition of parental rights in paradigm cases of procreation. Insofar as individuals have a presumptive right to control their reproductive gametes, they will always stand in a privileged position to form legitimate parental projects with respect to children who derive from those gametes, though it is ultimately their projects, rather than the fact of biological connectedness, that drives the rights-claim.

If the preceding remarks are sound, the parental project view presents three main advantages over Gheaus's gestational account, corresponding to its three major problems. First, it has a clear explanatory mechanism: whereas the gestational account must rely on controversial claims about maternal-fetal bonding, the parental project view relies on a widely accepted principle concerning non-interference in one's justice-respecting projects. This principle may of course be disputed, though not without enormous cost. If we were to reject the principle that all individuals have rights to non-interference in their justice-respecting projects, then we would be opening the door to the legitimate usurpation of our life projects by anyone who happens to develop an interest in them, whose ability to continue those projects would in turn be vulnerable to the legitimate usurpation of others, and so on *ad infinitum*. This would create a moral environment in which it was virtually impossible to formulate and pursue one's own conception of the good and would thus run counter to the fundamental commitment to individual autonomy that undergirds liberal theories of justice.

Second, the parental project view also escapes the inclusivity dilemma, as it is able to account for the rights of non-gestational partners without opening the door to a host of other potential rights-claimants as well. This is not only attractive insofar as it restricts legitimate parental rights-claims to a more plausible constituency of persons, but also because it allows for the adjudication of cases in which there are multiple conflicting claims over who gets to parent a newborn child. In the case of *Gestational Surrogate*, for example, the parental project view clearly and unambiguously assigns parental rights to the commissioning parents (Alice and Bill), whereas the relationship-based gestational account is ambiguous between assigning parental rights to the gestator (Claire) and the commissioning parents, assuming the commissioning parents have also participated in the various relationship-building activities that give rise to parental rights on that account. In this sense, the parental project view has the twin advantage of solving the baby redistribution problem *and* resolving cases in which there are multiple conflicting claims over who gets to parent a particular baby.

Finally, the parental project view can also explain our intuitions about important test cases where the gestational account goes astray. In the case of *Gamete Thief*, for example, the parental project view offers a clear explanation for why Ernest and Fran, rather than Dana, should be entitled to parent the child they deliberately set out to create. It also fares better in cases of gestation that do not involve a relational component. For example, the parental project view is able to explain why a woman who suffers from antenatal depression over the course of her pregnancy should nevertheless be entitled to parent her child, and it may also provide us with guidance in possible future cases of ectogenesis via artificial wombs, where a relationship-based gestational account would presumably be silent.

Despite these advantages, however, the parental project view also comes with three important caveats. First, because it is based on a more general right to non-interference in one's justice-respecting projects, the parental project view will not support the parental rights-claims of persons who have not formed parental intentions with respect to their progeny, including estranged biological parents. There have been numerous legal cases in which men, previously unknown to their biological children, have later claimed rights to parent those children on the basis of a genetic relationship.[50] The parental project view does not necessarily support such a rights-claim, as the assumption of a parental role with respect to those children cannot plausibly be interpreted as the continuation of a pre-existing project. In cases where a man was aware of his partner's pregnancy, but did not form parental intentions with respect to the child, he can be understood to have forfeited his initial opportunity to do so; similarly, in cases where a man was unaware of his partner's pregnancy, he could not have formed any parental intentions to begin with, and thus does not stand in any special rights position with respect to the resultant child. Of course, in cases where this information has been deliberately withheld, it is an open question whether a man has been wronged by being denied an initial opportunity to engage in the process of legitimate intention formation with respect to his progeny. The parental project view does not by itself supply an argument for such a wrong, though if such an argument was successful, the parental project view would have the resources to designate the partner's parental project as non-justice-respecting, and thereby close a moral loophole wherein women can deliberately withhold information about their pregnancies and exclude their male partners from being in a position to acquire parental rights.

Norvin Richards has recently supplied such an argument, claiming that a man may be wronged by having information about a pregnancy withheld from him if he had a legitimate expectation—rooted either in an explicit promise or the relationship's history—to believe that he would be included in the decision-making process following a conception.[51] If this is correct, then adhering to the justice-respecting condition of the parental project view may require his partner to offer him the opportunity to be included in her parental project. Of course, if this is incorrect, or if the case is one in which the man has no legitimate expectation to be included in the decision-making process following a conception, then the parental project view may well permit the legitimate exclusion of biological fathers in certain cases. One might object that this violates the Parity Principle—which I suggested earlier we have some reason to endorse on the grounds that (a) men and women have an equal interest in parenting, and (b) we are not responsible for our biological childbearing capacities—though it is not clear that it actually does. After all, the fact by which the woman is laying claim to being a parent in this case—her right to continue a justice-respecting project—is also a fact by which a man with equal merit could also lay claim to being a parent in many other cases. However, because pregnancies only occur in women's bodies (at least for now), it is sometimes the case that men will be in a weaker position to lay this claim.[52,53]

A second and related set of challenges involves cases in which one party forms parental intentions earlier than another and then seeks to exclude them from a parenting project. Imagine, for example, that a couple has conceived and is unsure of whether to proceed with the pregnancy. The man continues in his uncertainty, while the woman begins to envision a future in which she raises the child independently. Having now formed parental intentions with respect to the embryo she is carrying, does the woman have a right to non-interference against the man in what she now regards as *her* parental project? This, it seems, will again depend on whether or not the initiation of this project can be considered justice-respecting. The nature of the relationship might again play an important role in our thinking about this question. For example, if the man and woman are in a long-term relationship, the man may have grounds for a legitimate expectation to be included in the decision-making process following a conception, in which case the initiation of the woman's project would appear to be morally tainted. If, however, the conception is the result of a one-night stand, there may be no grounds for such an expectation, in which case it might not be unreasonable to think that the woman could legitimately form exclusionary parental intentions. In either case, custodial conflicts between biological progenitors represent an important set of challenges for the parental project view to address.

A third caveat is that in addition to claims about legitimate gamete access or expectations to be included in the decision-making process following a conception, the "justice-respecting" condition of the parental project view will also include claims about justice in child-rearing more generally. This means that in order for the parental project view to *unconditionally* support a right to parent one's biological child, we would have to make the following assumptions about the requirements of justice in child-rearing:

1 *Children may be permissibly raised by parents in families*: justice in child-rearing does not require (though it may permit) the non-parental care of children.
2 *Children may be permissibly raised by suboptimal competent parents*: justice in child-rearing does not require that children be allocated to the best available parents.
3 *The opportunity to parent is not a distribuendum of justice*: justice in child-rearing does not require an equitable distribution of parenting opportunities, which may otherwise be disrupted by automatically assigning parental rights to a child's biological parents.

Assumptions (1) and (2) are supported by the reasoning that underlies dual-interest accounts of parental rights: the protection of children's interests justifies assigning a small number of adults decision-making authority over a child, and, assuming those adults meet a minimum threshold of parental competency, their strong interest in exercising that authority justifies a right to do so without interference, even if there other adults available who would do a better job.[54] Notice, however, that the reasoning that gives plausibility to (1) and (2) also seems to render (3) indefensible. Indeed, if parenting contributes to human flourishing in the way that Brighouse and Swift (and others) suggest, then we ought to lament the situation of persons like George from *Baby Snatcher*, who are unable to satisfy their parenting interests due solely to brute bad luck. George's situation, while deliberately exaggerated, is structurally non-unique: single-sex couples and single or infertile individuals face similar unchosen barriers to the satisfaction of their parenting interests, particularly when there are no children readily available for adoption. Thus, the parental project view is best understood as supporting a right to parent one's biological child only under certain distributive conditions, namely, those of *baby surplus*, in which the number of children available to be parented exceeds the number of parenting units[55] who wish to become parents. In conditions of *baby scarcity*, in which the number of children available to be parented is less than the number of parenting units who wish to be parents, it is not unreasonable to think that the parental rights-claims of procreators might be tempered by the interests of disadvantaged prospective parents. If the opportunity to parent is itself a distribuendum of justice, then such persons may have a legitimate justice-based complaint in a scenario where heterosexual fertile couples were able to parent multiple children while *their* parenting interests went unsatisfied. In this type of scenario, achieving a more just distribution of parenting opportunities might require taking a number of measures currently viewed as supererogatory, including making gametes available for use in other people's procreative projects or even opening up the nuclear family to more multi-parenting arrangements.[56]

Notes

1 This view is typified by Tim Bayne and Avery Kolers, for example, who assume that "insofar as parenthood brings rights and responsibilities, it brings them *together*—that is, one does not get all the rights and none of the responsibilities, and vice versa." See Tim Bayne and Avery Kolers. "Towards a Pluralist Account of Parenthood," *Bioethics*, 17(3) (2003), pp. 221–242, 223.

2 The term "parental package" was coined by David Archard in reference to the view—which he rejects—that "both parental obligations and rights with respect to any particular child somehow come together as a whole 'package'." See David Archard. "The Obligations and Responsibilities of Parenthood" in David Archard and David Benatar (eds.) *Procreation and Parenthood: The Ethics of Bearing and Rearing Children* (Oxford: Oxford University Press, 2010), 106.

3 Anca Gheaus. "The Right to Parent One's Biological Baby," *Journal of Political Philosophy*, 20(4) (2012), pp. 432–455; and "Biological Parenthood: Gestational, Not Genetic," *Australasian Journal of Philosophy*, 96(2) (2018), pp. 225–240.

4 The following typology draws on similar typologies that have been advanced in the literature. See, for example, David Archard. *The Family: A Liberal Defense* (New York: Palgrave MacMillan, 2010), chapter 2, esp. at 28–29; Harry Brighouse and Adam Swift. *Family Values: The Ethics of Parent-Child Relationships* (Princeton: Princeton University Press, 2014), esp. 54–55; and Sarah Hannan. "Introduction" in Sarah Hannan, Samantha Brennan, and Richard Vernon (eds.) *Permissible Progeny? The Morality of Procreation and Parenting* (New York: Oxford University Press, 2015), esp. 12–16.

5 Hannan, "Introduction," 13.

6 Hannan, "Introduction," 14.

7 For child-centered approaches toward the justification of parental rights, see Archard, *The Family: A Liberal Defense*, esp. chapter 2; Jeffrey Blustein. *Parents and Children: The Ethics of the Family* (New York: Oxford University Press, 1982); Samantha Brennan and Robert Noggle. "The Moral Status of Children: Children's Rights, Parents' Rights, and Family Justice," *Social Theory*

and Practice, 23(1) (1997), pp. 1–26; Robert Noggle. "Special Agents: Children's Autonomy and Parental Authority" in David Archard and Colin Macleod (eds.) *The Moral and Political Status of Children* (Oxford: Oxford University Press, 2002); Sarah Hannan and Richard Vernon. "Parental Rights: A Role-Based Approach," *Theory and Research in Education*, 6(2) (2008), pp. 173–189; and Peter Vallentyne. "The Rights and Duties of Child-rearing," *William and Mary Bill of Rights Journal*, 11(3) (2003), pp. 991–1009.

8 For dual-interest approaches toward the justification of parental rights, see Harry Brighouse and Adam Swift. "Parents' Rights and the Value of the Family," *Ethics*, 117(1) (2006), pp. 80–108, as well as *Family Values*, esp. chapter 4; Matthew Clayton. *Justice and Legitimacy in Upbringing* (Oxford: Oxford University Press, 2006), chapter 2; Colin Macleod. "Parental Responsibilities in an Unjust World" in David Archard and David Benatar (eds.) *Procreation and Parenthood: The Ethics of Bearing and Rearing Children* (Oxford: Oxford University Press, 2010), and "Parental Competency and the Right to Parent" in Hannan, Brennan and Vernon, *Permissible Progeny?*; and Ferdinand Schoeman. "Rights of Children, Rights of Parents, and the Moral Basis of the Family," *Ethics*, 91(1) (1980), pp. 6–19.

9 Brennan and Noggle, "Children's Rights, Parents' Rights, and Family Justice," 4.

10 Brighouse and Swift, "Parents' Rights and the Value of the Family," 86.

11 Anne L. Alstott discusses the developmental importance of continuity of care in *No Exit: What Parents Owe Their Children and What Society Owes Parents* (Oxford: Oxford University Press, 2004), esp. chapters 1 and 2.

12 J. David Velleman. "Family History," *Philosophical Papers*, 34(3) (2005), pp. 357–378, and Bernard Prusak. *Parental Obligations and Bioethics: The Duties of a Creator* (New York: Routledge, 2013), esp. chapter 1.

13 For a persuasive account of these reasons, see Sally Haslanger's critique of Velleman in "Family, Ancestry and Self: What Is the Moral Significance of Biological Ties?" in *Resisting Reality: Social Construction and Social Critique* (Oxford: Oxford University Press, 2012), chapter 5.

14 While I focus here on Brighouse and Swift's intimacy-based account of this interest, it is important to note that it is not the only account available in the literature. On the contrary, Colin Macleod and Matthew Clayton have put forth alternative accounts based on, respectively, our interest in creative self-extension and pursuing our own conception of the good. See Macleod, "Parental Competency and the Right to Parent," 229–236, and Clayton, *Justice and Legitimacy in Upbringing*, 54–57.

15 See Brighouse and Swift, "Parents' Rights and the Value of the Family," 91–96, and *Family Values*, 87–93.

16 Brighouse and Swift, "Parents' Rights and the Value of the Family," 93.

17 Brighouse and Swift, "Parents' Rights and the Value of the Family," 95.

18 Gheaus, "The Right to Parent One's Biological Baby," 438–440.

19 As Brighouse and Swift put it, "Everything we have argued would apply in a world in which children were produced by storks." See Brighouse and Swift, *Family Values*, 104.

20 Gheaus, "The Right to Parent One's Biological Baby," 445–446.

21 It is worth noting, however, that not all philosophers have shied away from this option. Peter Vallentyne is happy to concede on child-centered grounds that the right to parent a newborn baby may be "legitimately claimed by anyone for whom possession is suitably in the child's best interests" (991), while Gheaus has defended the view that children should be allocated to the best available

parent. See Vallentyne, "The Rights and Duties of Child-rearing," and Anca Gheaus. "The Best Available Parent," *Ethics*, 131(3) (2021), pp. 431–459.

22 According to the Lockean theory of acquisition, "every man has a property in his own person: this no body has a right to but himself. The labour of his body, and the work of his hands, we may say, are properly his. Whatsoever then he removes out of the state that nature hath provided, and left it in, he hath mixed his labour with, and joined to it something that is his own, and thereby makes it his property." While Locke did not endorse the idea that procreators own their children, many theorists have suggested that this is an unavoidable implication of his theory of property acquisition. For Locke's theory of acquisition, see John Locke. *Second Treatise of Government*. Ed. C.B. Macpherson (Indianapolis: Hackett, 1980), chapter 5. For discussion of its implications in the case of procreation, see David Archard. *Children: Rights and Childhood* (New York: Routledge, 2004), 141–145; Robert Nozick. *Anarchy, State, and Utopia* (New York: Basic Books, 1974), 287–289; Susan Moller Okin. *Justice, Gender, and the Family* (New York: Basic Books, 1989), 74–88; and Hillel Steiner. *An Essay on Rights* (Oxford: Blackwell, 1994), 237–248.

23 Gheaus, 2012, 116–119. For a further discussion of the costs of pregnancy on which Gheaus's account explicitly draws, see also Amy Mullin. *Reconceiving Pregnancy and Childcare: Ethics, Experience, and Reproductive Labor* (Cambridge: Cambridge University Press, 2005).

24 Joseph Millum. *The Moral Foundations of Parenthood* (Oxford: Oxford University Press, 2016), 24.

25 There are also further problems with an account of parental rights based on a principle of investment. This account has plausibility when invoked as a principle to adjudicate custody disputes between adults who already stand in a parental role with respect to a child (e.g., following a divorce), though it is less plausible as a general account of how parental rights are initially acquired, as being in the appropriate rights position with respect to child is often a necessary precondition of performing what Millum takes to be rights-generating parental work. In this sense, the crucial question that needs answering in the present context—how one becomes initially entitled to perform parental work with respect to a particular child—is left unanswered by Millum's account.

26 Gheaus, "The Right to Parent One's Biological Baby," 449.

27 Gheaus, "The Right to Parent One's Biological Baby," 449.

28 Gheaus, "The Right to Parent One's Biological Baby," 451.

29 In the United States, precedent for assigning legal motherhood to the gestational mother was set in the famous 'Baby M' case, *In re Baby M*, 537 A.2d 1227, 109 N.J. 396 (N.J. 1988). In the United Kingdom, giving legal priority to gestational mothers is legislated through s. 33(1) of the *Human Fertilization and Embryology Act* (2008), which states that "The woman who is carrying or has carried a child as a result of the placing in her of an embryo or of sperm and eggs, and no other woman, is to be treated as the mother of a child."

30 Gheaus, "The Right to Parent One's Biological Baby," 452.

31 Lindsey Porter. "Gestation and Parental Rights: Why Is Good Enough Good Enough?" *Feminist Philosophical Quarterly*, 1(1) (2015), pp. 1–27, esp. 17–23.

32 Porter, "Gestation and Parental Rights," 22.

33 For example, I may have a strong attachment to the oak tree under which I proposed to my wife, though the bare fact that I have such an attitude is not in itself sufficient to confer rights to or over that tree, particularly if there are competing interests in the tree that must be balanced against my own (e.g., those of the community or other individuals).

34 Gheaus in fact hints at this strategy in Gheaus, "The Right to Parent One's Biological Baby," 452, n. 46, though it is potentially undermined by the empirical literature on maternal-infant bonding. Numerous studies suggest that the bond that develops between infants and their adoptive mothers is normally just as strong as the bond that develops between infants and their biological mothers, provided the infants are adopted shortly after birth. For two such studies, see Leslie M. Singer, David M. Brodzinsky, Douglas Ramsay, Mary Steir, and Everett Waters. "Mother-Infant Attachment in Adoptive Families," *Child Development*, 56(6) (1985), pp. 1543–1551; and J.E. Koepke, S. Anglin, J. Austin, and J. Delesalle. "Becoming Parents: Feelings of Adoptive Mothers," *Pediatric Nursing*, 17(4) (1991), pp. 333–336.

35 This type of account is of course vulnerable to a separate set of objections to which a relationship-based gestational account may be immune, including objections directed at the 'investment principle' expressed in (a).

36 Barbara Katz Rothman. *Reconceiving Motherhood: Ideology and Technology in a Patriarchal Society* (New York: W.W. Norton, 1989), 257.

37 David Archard and David Benatar. "Introduction" in David Archard and David Benatar (eds.) *Procreation and Parenthood: The Ethics of Bearing and Rearing Children* (Oxford: Oxford University Press, 2010), 26. The term 'Parity Principle' was originally coined by Tim Bayne and Avery Kolers in "'Are You My Mommy?' On the Genetic Basis of Parenthood," *Journal of Applied Philosophy*, 18(3) (2001), pp. 273–285, 280.

38 Gheaus, "The Right to Parent One's Biological Baby," 450.

39 Porter also makes this point in "Gestation and Parental Rights," 14–17.

40 It is also practically infeasible to the extent that the right to parent a particular child entitles the right-holder to a certain degree of exclusivity with respect to that child.

41 In Joseph Raz's widely cited version, "X has a right if and only if X can have rights, and, other things being equal, an aspect of X's well-being (his interest) is a sufficient reason for holding some other person(s) to be under a duty." See Joseph Raz. *The Morality of Freedom* (Oxford: Clarendon Press, 1986), 166.

42 Note that I do not assume that what drives the parental rights claim in this instance is the fact that the resultant child shares Alice and Bill's genetic material. The same reasoning would apply if the embryo was composed out of others' genetic material to which they had legitimate access, e.g., that of family members or a close pair of friends.

43 For example, while Dana intentionally steals an embryo from the outset, Claire initially engages in a legitimate agreement and only "steals" the fetus once she changes her mind about relinquishing it to Alice and Bill. Moreover, whereas Dana only steals the embryo, Claire steals the embryo *and* breaks a promise, giving Alice and Bill an additional moral complaint against her. However, neither of these differences detracts from the relevant fact that, in both cases, a person is claiming parental rights over an embryo that was created out of others' genetic material for the purpose of serving their parenting interest.

44 Gheaus defends a similar child-centered position in "Biological Parenthood," especially at pp. 236–237.

45 This would seem to imply—implausibly—that non-biologically related adoptive parents are inadequate, at least in cases where the gestator would otherwise meet the minimum threshold of parental competency.

46 One might wonder if it matters whether Ernest and Fran have multiple embryos stored, such that they would not be forever denied the opportunity to parent a

biological child if parental rights were assigned to Dana on child-centered grounds, even if they are forever denied the opportunity to parent this biological child. It is clear that they would have a much stronger complaint if the stolen embryo represented their only chance to pursue biological parenthood, though it does not follow that they would lack a legitimate complaint if they could "try again" with a different embryo. Whether or not this is the case will depend on a number of additional factors, including the extent to which we think the embryos are interchangeable and whether we think that the costs of separation to the child would outweigh the costs of undertaking another procreative project to Ernest and Fran. Furthermore, and perhaps more importantly, if Ernest and Fran's parental rights claim can be diminished on the grounds that they can "try again" with a different embryo, then presumably Dana's parental rights claim would also be diminished to the extent she can "try again" using an embryo to which she has legitimate access. In this sense, appealing to the ability to "try again" as a tiebreaker principle would not necessarily weigh in favor of assigning parental rights to Dana.

47 One important exception is found in the work of Norvin Richards, who sees the right to parent one's biological child as being rooted in a more general right to continue with whatever we have underway, so long as it does not cause harm to others. The account proposed here draws inspiration from Richards' work, though parts company with his account in a number of important ways, particularly in terms of the criteria it takes to be relevant for determining whether a project is in fact harmless (or, in my preferred terminology, justice-respecting). Nevertheless, it sits squarely in the same family of liberal views and may be read as a friendly complement to his account. For statements of Richards' account, see *The Ethics of Parenthood* (Oxford: Oxford University Press, 2010), esp. chapters 1–3, and "How We Acquire Parental Rights" in Leslie Francis (ed.) *The Oxford Handbook of Reproductive Ethics* (Oxford: Oxford University Press, 2016).

48 If we think that the opportunity to parent is itself a distribuendum of justice, we open up the possibility that the continuation of a parental project with respect to one's biological child may in fact infringe on the justice-based entitlements of other people, particularly those who are at an unchosen disadvantage in satisfying their own parenting interests (e.g., single, infertile, or homosexual individuals). I consider this possibility in further detail below.

49 While some philosophers derive this right from a more general right to bodily self-ownership (see, for example, Barbara Hall, "The Origins of Parental Rights," *Public Affairs Quarterly*, 13(1) (1999), pp. 73–82), it may also be derived from other rights that are less controversial. For example, if, in addition to the right to procreate, we think there is also a right *not* to procreate, the right to control the use of one's reproductive gametes can be plausibly derived from the right not to procreate.

50 For a widely discussed US case, see In the *Interests of B.G.C.*, Supreme Court of Iowa, No. 207/91-476, 92–49, September 23, 1992.

51 Richards, "How We Acquire Parental Rights," 273–279.

52 Some will undoubtedly find this to be an objectionable implication, but I think this reaction should be resisted for at least two reasons. First, if Richards' account is correct, then men may be able to guard themselves against this type of legitimate exclusion by only having sex with women who agree to include them in the decision-making process following a conception. Second, when viewed in the context of a complete account of parental rights and obligations,

the prospect of legitimate exclusion might be quite plausible. For example, those who find the prospect of legitimate exclusion *prima facie* objectionable due to the burden placed on men may be more sympathetic if, in excluding her male partner from acquiring parental rights, a woman can also be understood to have waived any legitimate claim for future assistance from the excluded man.

53 Another reason why men are in a weaker position to lay this claim is because they do not have the right to control whether a child will in fact be brought to term. Imagine that a woman has become pregnant with her partner, but is unsure of whether she wants to bring the baby to term. Her partner may form genuine parental intentions with respect to the embryo she is carrying, though this "project" cannot be considered justice-respecting insofar as it is conditional on the woman waiving her right not to procreate. There is a more general principle underlying this reasoning, which is that we cannot have a legitimate claim to non-interference in projects that require the potential interferer to waive their basic rights. For example, it is part of my conception of the good to grow old with my wife, and this imposes a duty on others not to interfere in our relationship in a number of ways. However, it does not impose a duty on my wife to stay in a marriage with me against her will, as she retains the right to free association and to set the terms of her own relationships.

54 This is not to suggest that either of these assumptions is uncontroversial. Gheaus herself has notably rejected (1) on the grounds that children have a right against being exposed to potentially harmful monopolies of care, while some philosophers have rejected (2) on the grounds that children have a right to be allocated to the best available parent upon birth. For Gheaus's rejection of (1), see "Arguments for the Nonparental Care of Children," *Social Theory and Practice*, 37(3) (2011), pp. 483–509, and "Children's Vulnerability and Legitimate Authority over Children," *Journal of Applied Philosophy*, 35(S1) (2017), pp. 60–75. For defenses of the best available parent view, see Vallentyne, "The Rights and Duties of Child-rearing"; Gheaus, "The Best Available Parent."

55 A parenting unit may be an individual or a small group of adults (e.g., a couple). A couple's parenting interests may be satisfied by jointly parenting one child between them; thus, in a surplus scenario, the number of children available to be parented need not exceed the number of individual adults who wish to be parents, as some of those adults will form collective parenting units.

56 For a discussion of the latter, see Anca Gheaus. "More Co-Parents, Fewer Children: Multiparenting and Sustainable Population," *Essays in Philosophy*, 20(1) (2019), pp. 1–21.

4 Procreation and Parental Obligations

4.1 Introduction

In the previous chapter, I argued that procreators have a presumptive moral right to parent the children they bring into existence insofar as the assumption of a parental role with respect to those children can be understood as the continuation of a justice-respecting parental project. In this chapter, I shift my focus from parental rights to parental obligations and ask whether procreators also have a presumptive moral obligation to parent the children they bring into existence. One way to get clearer on the distinction between these two questions is to consider the different contexts in which they typically arise. Questions about the origins of parental rights typically arise in cases where there is disagreement between different parties about who is *entitled* to parent a particular child, such as the surrogacy dispute cases discussed in the previous chapter. Questions about the origins of parental obligations, by contrast, typically arise in cases where there is disagreement between different parties about who is *responsible* for parenting a particular child. Consider, for example, one such case:

> *Unwilling Father*: Alan and Brenda meet for a casual sexual encounter, and both make it clear that they have no interest in establishing a long-term relationship and certainly no interest in having children together. However, despite taking every reasonable precaution to avoid conception, Brenda becomes pregnant and, having changed her mind about becoming a parent, decides to give birth to the child against her prior agreement with Alan. Is Alan now morally obligated to parent or otherwise support the child?

In order to adjudicate this type of case, we need a theory of parental obligation, or a theory about *how*, or in light of which facts, one acquires an obligation to parent a particular child. The existing literature on the morality of parenthood offers three main theories, which are normally presented

DOI: 10.4324/9781003164104-5

as rivals: the *genetic* account, which holds that one acquires an obligation to parent a child by virtue of contributing the genetic material from which that child developed[1]; the *voluntarist* account, which holds that one acquires an obligation to parent a child by consenting to do so[2]; and the *causal* account, which holds that one acquires an obligation to parent a child by causing that child to exist and hence to be in need of parental care.[3] Each theory presents a number of advantages and disadvantages in their ability to account for parental obligations in a variety of cases. For example, the genetic account is consistent with the way that parental obligations are typically assigned in the law and can support the judgment that an unintentional father like Alan is obligated to parent his genetically related child. However, it also has counterintuitive implications when applied in other types of cases (e.g., gamete theft) and appears to lack an explanatory basis, for it is unclear how the mere fact of genetic connectedness can give rise to weighty moral obligations. By contrast, the voluntarist account provides a straightforward explanation of how parental obligations are incurred in standard cases of intentional procreation, and one that is consistent with voluntarist accounts of role obligations more generally. However, it is unable to account for parental obligations in cases where procreators explicitly refuse to undertake them. This not only leads to the conclusion that Alan has no obligation to parent the child he created with Brenda—which proponents of voluntarism generally find acceptable—but also to the more troubling conclusion that there may be children born who no one is uniquely obligated to parent. Finally, the causal account appears to avoid this implication by assigning parental obligations on the basis of one's causal relationship to a child's existence, but in so doing invites a host of normative and conceptual problems that appear to undermine it as a tenable account. For instance, critics have argued that the causal account over-determines who counts as parentally obligated; that it can only account for limited compensatory obligations; that it is inconsistent with a standard fault-based view of liability; and that it is unfair by assigning equal parental liability to parties who normally have unequal decision-making power over whether or not a child is caused to exist. The combination of these problems has presented a serious challenge to proponents of the causal account, leading critics like Teresa Baron to conclude that causal theories of parental obligation are ultimately a "lost cause."[4]

In this chapter, I will argue that this conclusion is premature. Not only can a suitably modified version of the causal account accommodate the multitude of objections that have been raised against it, but the purportedly attractive features of the voluntarist account—including its capacity to not hold involuntary fathers liable in some cases and to account for the full range of duties that parents intuitively owe to their children—can also be fully accommodated within a causal framework. The relevant modifications consist in clarifying the principle of moral responsibility that underlies the causal account as well as the nature of the need that is created as a result of causing a child to exist. Both of these elements have been mis- and/or under-described by proponents of the causal account in ways that create significant vulnerabilities in their accounts, though these vulnerabilities are not endemic to causal accounts and may be overcome by a suitably amended theory. In this chapter, I outline and defend one such variation, which I term the *moral responsibility for parental need* account of parental obligation. According to this account, being morally responsible for a child's need for long-term parental care is sufficient to generate an obligation to ensure that the child is adequately parented, which may be discharged by either parenting the child oneself or by delegating caretaking responsibilities to a capable and willing volunteer. On this account, whether or not Alan is obligated to parent the child created out of his sexual encounter with Brenda will depend on whether he satisfies key conditions for the attribution of moral responsibility as well as whether assuming a parental role with respect to the child is necessary to ensure that the child is adequately parented, variables that I will argue are importantly context dependent.

The chapter proceeds in four parts. In Section 4.2, I outline the three dominant theories of parental obligation and explain why only the causal account is capable of establishing a necessary moral relationship between procreation and the acquisition of parental obligations. In Section 4.3, I defend a particular version of the causal account by outlining and responding to five major challenges that have been raised against causal accounts in general. In Section 4.4, I summarize the main tenets of this particular version of the causal account as the *moral responsibility for parental need* account of parental obligation. Finally, in Section 4.5, I offer some general conclusions about the moral relationship between procreation and parenthood and outline some of its normative implications.

4.2 Theories of Parental Obligation

Consider the following case:

> *Willing Procreators*: Charles and Diane are in a stable long-term relationship and have long dreamed of parenting a child together. Deciding the time is now right, Diane stops taking her birth control pills and has unprotected sex with Charles at the times of the month when she is most likely to conceive. Shortly thereafter, Diane becomes pregnant, and nine months later gives birth to a healthy child. She and Charles then take the baby home from the hospital and eagerly begin their new lives together as parents.

It is plausible to think that Charles and Diane are now morally obligated to parent their biological child. But what is the basis of this obligation? Existing theories of parental obligation give three different answers, which in turn have different implications for other cases of procreation. The *genetic* account locates the obligation in the fact that the child has developed from Charles and Diane's reproductive gametes; the *voluntarist* account locates the obligation in the fact that Charles and Diane appear through their actions to have tacitly consented to parent their child; and the *causal* account locates the obligation in the fact that Charles and Diane are causally responsible for their child's existence and hence need for parental care. Each of these theories may be posited in a *strong* or a *weak* sense, or as indicating a necessary or sufficient condition for the acquisition of parental obligations. Positing a strong version of any account commits one to *monism* about parental obligation, or the view that obligations to parent children can only be acquired in one way, whereas positing a weak version of any account is consistent with *pluralism* about parental obligation, or the view that obligations to parent children can be acquired in a variety of ways.[5] The view that I defend in this chapter is consistent with and in fact assumes a pluralistic approach toward parental obligation in which the obligation to parent a child can be acquired through being morally responsible for a child's need for long-term parental care and/or through voluntarily accepting a parental role with respect to an eligible child. However, before outlining and defending this particular view in greater detail, it is first necessary to introduce the existing views with which it competes.

4.2.1 *The Genetic Account*

According to the genetic account, one acquires an obligation to parent a child by contributing the genetic material from which that child developed. The material in question has historically been conceived in terms of reproductive gametes—sperm and ova—though with the recent advent of three-parent IVF, it may be extended to include the mitochondrial DNA contributed by nuclei donors as well.[6] On the genetic account, Charles and Diane are obligated to parent their biological child because they have contributed the genetic material from which that child developed. Importantly, this obligation is shared between them, as both contributed genetic components that were necessary for that child's existence.

If posited as a strong or monistic theory of parental obligation, the genetic account is clearly implausible, as it cannot account for the obligations of non-genetically related adoptive parents. However, even if genetic connectedness is not necessary for the acquisition of parental obligations, one might be tempted to think it is at least sufficient in many cases. One reason for this is that it appears to be consistent with the way that parental obligations are typically assigned in the law. In the United Kingdom, for example, the establishment of genetic paternity normally provides legal grounds for securing child support payments from men who have contributed genetic material to a child their sexual partner has brought to term, regardless of the nature of their relationship with that partner or their intentions in having sex. Similarly, common law in Australia has also relieved men of the obligation to financially support a child when paternity tests have revealed they are not in fact the child's genetic father.[7] To the extent that we endorse these types of judgments, we may be led to the view that genetic connectedness can generate parental obligations.

Despite its popular appeal and apparent instantiation in law, however, the genetic account faces at least two major problems that ultimately discredit it as a plausible theory of parental obligation. The first problem is that it leads to unacceptable implications when generalized to other types of cases. Consider, for example, one such case:

> *Sperm Heist*: Emma breaks into the freezer of a gamete storage facility and steals a sample of sperm that Fred had frozen before his vasectomy. She then uses it to inseminate herself, and nine months later gives birth to a healthy child.

According to the genetic account, Fred would now be obligated to parent the child in light of contributing part of the genetic material from which the child developed, yet holding Fred liable for this obligation seems egregiously unfair given that he did not consent to, and was not even aware of, his contribution. Proponents of the genetic account might seek to avoid this implication by adding the condition that the genetic material must be contributed willingly and in full view of the potential consequences, though even this qualified view would still imply that voluntary gamete donors are equally liable to support the children who develop from their reproductive gametes. Some theorists are happy to accept this implication as a defensible feature of an account of parental obligation,[8] though we should be hesitant to accept it if there are alternative theories that are capable of avoiding it.

The second problem with the genetic account has to do with its lack of explanatory power, or its inability to account for the relationship between genetic connectedness and the acquisition of moral obligations. A satisfactory account of the origins of parental obligations should be able to explain how a certain set of facts gives rise to a certain set of moral obligations, though as Joseph Millum puts it, "it is inexplicable what justificatory role stretches of DNA could play."[9] This lack of explanatory power is helpfully illustrated by considering how genetic connectedness functions (or fails to function) in non-procreative contexts. Suppose that I have an identical twin brother with whom I share all of my DNA and to whom I have special obligations. While there are numerous explanations that may be given for the origins of these obligations—our shared history, my past promises to him, his vulnerability to my actions—our genetic connectedness does not seem to be one of them. In fact, if genetic connectedness were sufficient for the acquisition of special obligations, it would imply that I would still have special obligations to my brother even if we were separated at birth and grew up in isolation from one another on opposite sides of the world. This implication is clearly false, however, suggesting that genetic connectedness alone is not sufficient for the acquisition of moral obligations.

A proponent of the genetic account might respond to these objections by claiming it is not genetic connectedness *per se* that explains how procreators become obligated to their biological progeny, but rather a principle pertaining to their responsibility for their reproductive gametes. For example, Rivka Weinberg argues that reproductive gametes are akin to a hazardous material, for if we are not careful with them, they have the potential to join with other reproductive gametes to create needy individuals with full moral status.[10] Being in possession of hazardous materials normally generates an extremely high standard of care, and an attendant responsibility to deal with the consequences when that standard is not upheld. Thus, Weinberg argues that "when we choose to engage in activities that put our gametes at risk of joining with others and growing into persons, we [must] assume the costs of that risky activity"[11] by undertaking the responsibility to care for those persons.

Weinberg's "Haz-Mat" theory of parental responsibility provides a much stronger explanation of how procreators become obligated to their genetic progeny, and one that may be able to avoid some of the counterintuitive implications of less sophisticated genetic accounts. For example, it will not assign parental obligations to victims of rape or gamete theft, as those persons have not voluntarily engaged in activities that put their gametes at risk of joining with others and growing into persons. Notice, however, that by invoking a principle of control over our reproductive gametes, the Haz-Mat theory ceases to function as a genetic account and simply collapses into a version of the causal account that assigns parental obligations on the basis of one's moral responsibility for a child's existence. Indeed, procreators on Weinberg's theory do not become obligated to their genetic progeny *because* they are genetically related, but rather because they have acted in ways—either intentionally or negligently—that have resulted in the creation of needy individuals with full moral status. Just as we may act in these ways using our own reproductive gametes, we may also do so using others' reproductive gametes over which we have control, as in the case of *Sperm Heist*. In this sense, if genetic connectedness is relevant for the acquisition of parental obligations—as it often appears to be in the law—it is only as a proxy for a more normatively relevant property, such as being morally responsible for the existence of a child in need of parental care.

4.2.2 *The Voluntarist Account*

A more promising alternative to the genetic account is premised on the notion of consent, or voluntary acceptance of the parental role. The voluntarist account follows from a more basic principle concerning the acquisition of role obligations, namely, the *volunteer principle*, which holds that "the only way in which we can acquire role obligations—or in any case, role obligations with genuine moral force—is by signing on for the roles to which they are attached."[12] Many critics have objected to the volunteer principle on the grounds that it is not universally applicable, as it cannot account for presumptively involuntary role obligations like filial obligations or obligations of citizenship.[13] However, even if the volunteer principle cannot account for all of the role obligations that we take to be important, it nevertheless specifies at least one common way in which role obligations are incurred. The role-based obligations of spouses, friends, and employees, for example, can all be explained in voluntarist terms: in each case, the obligations are typically incurred as a result of individuals having signed on for the roles to which they are attached.

In the context of parenthood, the volunteer principle implies that parental obligations are also incurred as a result of signing on for the role of parent. "Signing on" here may take an explicit form, as when one formally adopts a child through a legal procedure, though it may take an implicit form as well, as when one performs a set of actions that one knows, or ought to know, signals the assumption of the parental role, such as Charles and Diane deliberately creating a child and taking it home from the hospital upon its birth.[14] It is important to note, however, that while some form of voluntary acceptance is necessary for incurring parental obligations on a voluntarist account, it is not by itself sufficient, as candidate parents must normally meet at least two additional conditions of eligibility.[15] First, they must stand in the appropriate rights position with respect to a child in order to assume the role of that child's parent. If that role is already filled, or if someone else has a better claim to fill it, then simply volunteering to be a child's parent is not sufficient for acquiring parental obligations. Second, a candidate parent must also be capable of discharging the relevant obligations. If a person is demonstrably unfit to parent a child, then they may not undertake the role of a child's parent, even if no one else has a better claim to do so.

As an account of parental obligation, the voluntarist account has a number of advantages. First, unlike the genetic account, it is explanatorily strong: it offers a clear and plausible explanation of how parental obligations are acquired, and one that is consistent with generally accepted principles concerning the acquisition of role obligations. Second, it is also broadly inclusive in terms of who it can designate as parentally obligated. Whereas the genetic and causal accounts only seem capable of accounting for the obligations of persons who have caused a child to exist through procreation, the voluntarist account is capable of accounting for the parental obligations of both procreative and adoptive parents.

Despite these advantages, however, the voluntarist account also runs into a number of difficulties when invoked in concrete cases of procreation. First, while it has no trouble assigning parental obligations in standard cases of intentional procreation, it encounters a potential difficulty in cases where there is disagreement between procreators about who is responsible for parenting the resultant child. In the case of *Unwilling Father*, for example, the voluntarist account implies that Alan has no obligations to the child created out of his sexual encounter with Brenda, for not only did he take every reasonable precaution to avoid conception, but he also explicitly expressed his desire *not* to become a parent. Some proponents of voluntarism have seen this as an attractive feature of their account, for if women have the choice of opting out of their parental obligations through abortion or putting a child up for adoption, then fairness demands that men should have a similar choice as well.[16] Others, however, have seen this implication as a key inadequacy, and a potential *reductio ad absurdum* for voluntarism in general.[17]

Of course, whether or not we take this implication to be problematic will depend on our pre-reflective intuitions about cases like *Unwilling Father*, and not everyone shares the intuition that unwilling biological fathers are necessarily obligated to parent their unwanted progeny. However, a second and more serious problem for the voluntarist account is that it appears vulnerable to what Elizabeth Brake calls the *deadbeat dads problem*—which I will call the *child abandonment problem*, given that it applies equally to biological fathers and mothers—wherein procreators may permissibly abandon their offspring by not volunteering to serve as their offspring's parents. Judith Jarvis Thomson recognizes (and appears to endorse) this implication of voluntarism, claiming,

> Surely we do not have any such 'special responsibility' for a person unless we have assumed it, explicitly or implicitly. If a set of parents do not try to prevent a pregnancy, do not obtain an abortion, and then at the time of birth of the child do not put it out for adoption, but rather take it home with them, then they have assumed responsibility for it, they have given it rights, and they cannot *now* withdraw support from it at the cost of its life because they now find it difficult to go on

> providing for it. But if they have taken all reasonable precautions against having a child, they do not simply by virtue of their biological relationship to the child who comes into existence have a special responsibility for it. They may wish to assume responsibility for it, or they may wish not to. And I am suggesting that if assuming responsibility for it would require large sacrifices, they may refuse.[18]

Brake has argued that the child abandonment problem is in fact illusory, for even if voluntary acceptance is necessary for acquiring parental obligations, general moral duties will normally block the conclusion that procreators may permissibly abandon their offspring. As she claims, "if such abandonment endangers the child, the claim that a voluntarist account would permit it is false: it would be wrong for anyone to abandon a child to death, because it violates a general duty of easy rescue."[19] An appeal to general moral duties may indeed block the conclusion that procreators may abandon their offspring, though it is at best a partial response to the child abandonment problem. Indeed, what is troubling about this implication of voluntarism is not only the prospect of permissible infanticide but also the prospect that there may be children born whom no one is uniquely obligated to parent. It would be consistent with Brake's view, for example, for procreators to sustain their progeny at a subsistence level but to deny them the more extensive goods of care, nurturing, and intimacy that are associated with parent-child relationships. If we view such a scenario as an injustice to children, then the child abandonment problem remains, and the voluntarist account appears inadequate as a monistic theory of parental obligation.

4.2.3 The Causal Account

The causal account of parental obligation seeks to avoid the child abandonment problem by assigning parental obligations to persons on the basis of their causal responsibility for a child's existence and hence need for parental care. Whereas the voluntarist account follows from a more basic principle concerning the acquisition of role obligations, the causal account follows from a more basic principle of morality in general, namely, the principle that we ought to bear responsibility for the outcomes of our actions. Proponents of the causal account often defend their view by way of analogy with everyday cases in which causing a person to be in a position of need generates an obligation to ensure that need is fulfilled. As a typical example of such a case, consider David Velleman's analogy of a child who is pushed into the deep end of a swimming pool:

> [Imagine] a child who is drowning at the deep end of a swimming pool. People lounging around the pool obviously have an obligation to rescue

> the child. But the obligation to fish the child out doesn't fall on the bystanders equally if one of them pushed the child in. The one responsible for the child's predicament is not just a bystander like the others, and he bears the principal obligation ... Likewise with procreation and parenting. In my view, parents who throw a child into the predicament of human life have an obligation to lend the assistance it needs to cope with that predicament, by helping it to acquire the capacities whose exercise will enable it to flourish and whose lack would cause it to suffer. By choosing to create a child, perhaps even by choosing to have sex, adults take the chance of incurring this obligation.[20]

On this view, the obligation to parent a child one has created is not grounded in one's voluntary acceptance of the role, but rather in the fact that one has caused a child to exist in a dependent and vulnerable state where they now require long-term parental care. It does not matter, then, that Charles and Diane have tacitly consented to parent their child in the case of *Willing Procreators*. Even if they changed their minds upon the child's birth and no longer wanted to become parents, they would still be obligated to the child by virtue of having caused that child to be in need of parental care.[21]

It is important at this point to distinguish two possible interpretations of the relationship between causing a child's existence and acquiring parental obligations. On the first, standard interpretation, causing a child to exist *directly* obligates one to undertake the role of that child's parent. This is the interpretation reflected in Velleman's analogy where, much like the person who pushed the child into the pool, the persons responsible for a child's existence are directly responsible for lending the assistance it requires to cope with its newfound predicament. However, there is also a second interpretation in which causing a child's existence and acquiring parental obligations are *indirectly* related. On this interpretation, the obligation that is incurred as a result of causing a child to exist is not the obligation *to* parent the child, but rather the distinct and prior obligation to ensure *that* the child is adequately parented. This obligation may be and often is discharged by parenting the child oneself, though it may also be discharged by ensuring that someone else parents the child, as in standard cases of adoption. It is worth noting that this indirect version of the causal account is necessarily committed to pluralism about parental obligation, for it assumes that parental obligations can be incurred both through causing a child to exist and by voluntarily accepting a parental role, as in the case of adoptive parents who are enlisted to care for a child who others have caused to exist.

Within the indirect version of the causal account, it is also important to distinguish between two possible positions on the transferability of parental obligations. On the first view, which we may term the *transfer view*, the obligation to ensure that a child is adequately parented may be fully discharged by transferring custody of the child to one or more willing adoptive parents. In this case, the procreators have fulfilled their obligation to the child and it is then the adoptive parents' responsibility to ensure the child is adequately cared for from that point forward, e.g., by making custody arrangements for the child in the event of their death. On the second view, which we may term the *delegation view*, the obligation to ensure that a child is adequately parented is inalienable: while procreators may delegate caretaking responsibilities to one or more willing adoptive parents, they still retain an obligation to ensure that the child is adequately parented should the adoptive parents fail to fulfill those responsibilities.[22]

If we assume that adoptive parents are just as capable as biological parents of providing their children with a supportive upbringing,[23] then the indirect version of the causal account actually fits better with the analogy that Velleman uses to support the direct version. Presumably, pushing a child into a pool generates an obligation to ensure *that* the child is pulled out of it, though discharging this obligation does not necessitate pulling the child out oneself: if there are other persons available who are willing to pull the child out, and who would do just as good (or better) of a job as oneself, then it seems as though one may satisfy their obligation to the child by ensuring that someone else comes to his assistance. At the same time, however, it also seems that one would not escape their obligation to the child should the persons enlisted to rescue him fail in their task. In this case, the obligation generated as a result of causing the child's predicament would not have been fulfilled and thus ought to be attended to by the initial duty-bearer, i.e., the causally responsible agent. In what follows, I assume in accordance with the indirect version of the causal account that the obligation that is incurred as a result of causing a child to exist is not the obligation *to* parent the child, but rather the obligation to ensure *that* the child is adequately parented. I also assume in accordance with the delegation view that this obligation is inalienable: while procreators may discharge this obligation by delegating caretaking responsibilities to one or more adoptive parents, they still retain an obligation to ensure the child is adequately parented in the event that the adoptive parents fail to provide adequate parental care. As I will demonstrate below, keeping these distinctions in mind is key to responding to many of the standard objections that have been raised against the causal account.

Regardless of which version we endorse, however, the causal account in general presents a number of advantages as a theory of parental obligation. First, like the voluntarist account, the causal account is also explanatorily strong: it offers a clear and plausible explanation of how parental obligations are incurred, and one that is consistent with generally accepted principles concerning moral responsibility. Second, unlike the voluntarist account, the causal account is also capable of assigning parental obligations in cases where one or both procreators do not wish to parent a child they have caused to exist. In the case of *Unwilling Father*, for example, the causal account is consistent with (though, as I will argue in Section 4.3.4, does not necessarily imply) the view that Alan is obligated to parent the child created out of his sexual encounter with Brenda, for even though he expressed his intention not to become a parent, he nevertheless acted in a way that caused the existence of a vulnerable child in need of parental care. The same reasoning generalizes to cases in which both procreators do not wish to parent their child, thereby avoiding the child abandonment problem.

Despite these advantages, however, the causal account has also been the subject of considerable criticism in the literature surrounding the morality of parenthood. In addition to the obvious objection that it cannot account for the obligations of parents who do not cause their children to exist (which would only apply to strong or monistic versions), critics have also raised five more troubling objections, namely: (1) that the causal account over-determines who counts as parentally obligated; (2) that it can only account for limited compensatory obligations; (3) that it is inconsistent with a standard fault-based view of liability; (4) that it is unfair by assigning equal parental liability to parties who normally have unequal decision-making power over whether or not a child is caused to exist; and, regarding the indirect version specifically, (5) that it fails to register the moral significance of biological connectedness by permitting the delegation of caretaking responsibilities to non-biologically related adoptive parents. My objective in the remainder of this chapter is to demonstrate how a suitably modified version of the causal account can accommodate these worries and thus provide a compelling account of the moral relationship between procreation and the acquisition of parental obligations.

4.3 Defending the Causal Account of Parental Obligation

4.3.1 The Too Many Parents Problem

The first challenge for the causal account of parental obligation consists in providing a non-arbitrary conception of causation that does not over-determine the constituency of causally responsible agents. Causation is often conceived in counterfactual or "but-for" terms, whereby an action counts as a cause for an event if that event would not have occurred but-for that action being performed.[24] As several critics have noted, however, this conception of causation risks being far too inclusive, as it would seem to assign parental obligations to virtually any person who has participated in a causal chain that has resulted in a child's birth.[25] For example, in an

IVF pregnancy, there are multiple persons who causally contribute to the resultant child's existence, including the man and woman who supply the reproductive gametes, the embryologist who inseminates the egg, the clinician who performs the embryo transfer, the woman who gestates the embryo (who may be distinct from the woman who supplied the ovum), and the obstetrician who administers the pregnancy—not to mention the multitude of persons, including matchmakers, grandparents, and pronatalist friends, who contributed to the conditions under which the decision to procreate was initially made. A standard but-for conception of causation would seem to imply that each of these persons bears partial responsibility for the child they causally contributed to bringing into existence, leading to "a tremendous proliferation in the number of persons who could be considered [a] child's 'parents'."[26]

Some causal theorists have attempted to address this worry by separating the primary and proximate cause of a child's existence from the infinite chain of necessary conditions that preceded it. For example, Michael Austin has suggested that we can posit a

> common sense notion of causation which takes genetic (and perhaps gestational) parents to be the primary and proximate causes of the existence of their children and rules out those individuals who make important causal contributions to a child's existence, including healthcare professionals.[27]

This type of approach might succeed in narrowing down the range of causally responsible agents, though it seems to do so at the cost of circularity, for it simply defines as primary and proximate the causal contributions of the persons who we pre-reflectively consider to be obligated to the child, without actually explaining how *their* causal contributions are more relevant to the assignment of parental obligations than the causal contributions of possible others, e.g., the IVF clinician who also played a proximate causal role in the child's existence. This is clearly unsatisfactory. The very purpose of the causal account is to explain *who* bears responsibility for parenting a particular child, so we cannot very well appeal to those persons when determining a conception of causation.

A better approach to resolving the too many parents problem begins by distinguishing causal from moral responsibility. Notice that the too many parents problem is premised on the assumption that there is a direct and perhaps proportional correlation between causal influence and moral responsibility, such that causally contributing to a particular event is sufficient for bearing moral responsibility for it. This assumption is false in at least two respects. First, we do not normally hold people to be morally responsible for *all* of the remote consequences of their actions, even when those actions serve as necessary conditions for the consequences in question. For example, by pioneering the field of nuclear physics, Ernest Rutherford played a necessary causal role in the atomic explosions at Hiroshima and Nagasaki, though it would be a mistake to hold him morally responsible for those explosions in light of that role—after all, he did not participate in the construction or delivery of the bombs, nor was the creation of atomic bombs a foreseeable consequence of his research at the time it was being conducted. In this sense, while causally contributing to a particular event is normally necessary for bearing moral responsibility, it is not in itself sufficient.[28]

Critics like Austin will likely point out that the reason Rutherford is not morally responsible is that he is too far removed—while he played a necessary causal role in the explosions, he was not their primary and proximate cause. Notice, however, that this explanation is also incomplete, as we do not necessarily hold people to be responsible for every event for which they are the primary and proximate cause, at least not in the absence of additional conditions. If I light a cigarette in a hallway where there happens to be a gas leak, I am the primary and proximate cause of the subsequent explosion, but it does not necessarily follow that I am morally responsible for the destruction that ensues. For me to bear moral responsibility for the explosion, it seems like additional conditions have to be in place, e.g., that I acted intentionally, that the lighting of the cigarette was uncoerced, that I was aware of the risk of a gas leak (or was culpably ignorant of such a risk), and so on. Again, while causally contributing to an event appears to be necessary for bearing moral responsibility, it does not appear to be sufficient, even in cases where the causal contribution in question is the most primary and proximate.

What both of these examples illustrate is that the relationship between causation and moral responsibility is not direct or straightforward but is rather mediated by an additional set of considerations that serve to link the two concepts. The question we should be asking, then, is not the empirical or metaphysical one of who played a necessary causal role in a child's existence—a group that undeniably includes grandparents, matchmakers, fertility clinicians, gamete donors, obstetricians, etc.—but rather the normative one of what conditions have to obtain for a person's causal contribution to translate into moral responsibility. It is beyond the scope of this chapter to provide a full account of the relationship between causation and moral responsibility, though consider at least two basic conditions that many philosophers believe are individually necessary and jointly sufficient for an agent to bear moral responsibility for an outcome they have caused.[29] First, the agent in question must have a sufficient degree of control over the actions that lead to a particular outcome (call this the *control* condition). This implies, for example, that they were not acting under conditions of duress and that there were alternative courses of action available to them, such that they could have avoided a particular outcome by choosing to do otherwise. Second, the agent's mental state must also be such that they can properly be held accountable for the outcomes of their actions (call this the *epistemic* condition). This implies, for example, that they possess basic cognitive powers such as means-end reasoning, that they were acting intentionally, that they were aware (or ought to have been aware) that a particular course of action may result in a particular outcome, and that they were aware (or ought to have been aware) of alternative courses of action that could have avoided a particular outcome.

When one or both of these conditions are unsatisfied, it is normally inappropriate to hold an agent morally responsible for the outcomes of their actions. To take a simplified example, imagine that turning on a light switch will deliver a painful electrical shock to an unsuspecting victim in the next room. Even if I turn on the light switch and directly cause the electrical shock, I cannot be held morally responsible if I am forced to turn on the light switch or if I am unaware and had no reason to believe that doing so would cause an electrical shock to an unsuspecting victim. Notice, moreover, that both of these conditions also explain why causal proximity often seems relevant to the assignment of moral responsibility. We do not normally hold people to be morally responsible for the distant outcomes of causal sequences that spring from their actions, though this is not *because* those outcomes are distant, but rather because they are often insufficiently foreseeable at the time of the action in question and/or because they are dependent on further actions or interventions by others over which the original causal agent has insufficient control.[30]

To be sure, these two conditions provide only a partial picture of the complex relationship between causation and moral responsibility, and the conditions themselves could be subject to further interrogation—we could go into significantly more detail, for example, about what sufficient control entails or what epistemic criteria are most relevant for the assignment of moral responsibility. Even stated broadly, however, notice that these conditions go a long way toward narrowing the range of morally responsible causers to a much more plausible set. If we take control to be a necessary condition of moral responsibility, then parental obligations can only be incurred as a result of voluntary[31] actions that result in the existence of a child. This will rule out the assignment of parental obligations to victims of sexual assault, as well as to women who are compelled to bring accidental pregnancies to term (e.g., in cases where access to safe abortion is restricted). Moreover, if we take the aforementioned epistemic criteria to be necessary for moral responsibility, then parental obligations can only be incurred in cases where the creation of a child is a reasonably foreseeable consequence of a voluntary action performed by a competent agent. This will rule out the assignment of parental obligations to victims of deceptive sexual practices (e.g., lying about fertility or poking holes in condoms), as well as to persons who lack the cognitive powers to understand the potential consequences of their actions (e.g., the severely cognitively impaired). Finally, the combination of the control and epistemic conditions will normally rule out the assignment of parental obligations to persons like grandparents, matchmakers, and pro-natalist friends, for even though they may have contributed to the conditions under which a decision to procreate was initially made, the creation of a child is not always a foreseeable consequence of their particular causal contributions, nor is it one over which they have a sufficient degree of control to be held morally responsible. Indeed, setting up a date or professing the joys of parenthood does not typically result in the creation of a child, at least not in a way proximate enough to confer moral responsibility for the child's existence.

The constituency of morally responsible causers, then, is limited to competent agents who act freely in ways that can be reasonably foreseen to result in the creation of a child without the need for significant subsequent intervention on the part of other agents in a way that diminishes the level of control they have over the outcomes of their actions. This includes persons who freely engage in heterosexual intercourse regardless of whether they intend to create a child, though it also seems to include persons like fertility clinicians or gamete donors, as they too act freely in ways that can be reasonably foreseen to result in the creation of children.

How should a causal theorist deal with the inclusion of these latter groups? Lindsey Porter believes their inclusion is actually palatable on an indirect version of the causal account, for while persons like gamete donors and fertility clinicians may incur obligations to ensure *that* the children they cause to exist are adequately parented, these obligations are normally discharged by ensuring that persons who access their services assume parental care of those children. It ought to be noted, however, that as a fellow proponent of the delegation view, Porter believes that these obligations are inalienable, such that gamete donors and fertility clinicians *may* incur obligations to parent the children they cause to exist in the event that their clients fail to discharge their parental responsibilities. Several commentators have regarded this implication as deeply implausible and hence as an indictment of Porter's approach in general.[32]

A better and much simpler approach for dealing with gamete donors, fertility clinicians, and any other person who offers essential assistance in the creation of children (e.g., surrogates, obstetricians, or midwives) is to regard those persons as representatives acting on other agents' behalf. Jason Hanna introduces the analogy of employment to capture this type of relationship. According to Hanna,

> when an agent, A, acts in a responsible and non-negligent way on behalf of another, B, in the context of a relationship such as that between an employee and employer, then special obligations resulting from A's behavior accrue to B, rather than A.[33]

To take Hanna's own example, imagine that I am contracted by a mining company to dig an underground mine, which involves the release of pollutants that may harm local residents if they are not cleaned up within a year. Suppose further that the mining company has assured me it will clean up the pollutants within this time frame and that I have every reason to believe their assurances. Even though my actions have directly caused the release of the pollutants, it seems clear that I have no personal liability to assume the cost of the cleanup insofar as I was acting non-negligently as a representative of the mining company, and this remains true even if the mining company later reneges on their commitment and decides it is either unable or unwilling to assume the cost of the cleanup. Notice, however, that our judgments would change if I acted negligently in my capacity as a representative. For example, imagine that I knew or ought to have known that the mining company would not in fact clean up the pollutants, but still decided to proceed with digging the mine. In this case, it would not be unreasonable to suggest that I may be personally liable for the cost of the cleanup and/or for compensation owed to local residents who may have been harmed by the release of the pollutants.

Parallel reasoning applies to gamete donors, fertility clinicians, and other medical practitioners who offer essential assistance in the creation of children. Rather than thinking of these persons as independent agents whose causal contributions make them liable to incur parental obligations, we ought to regard them as representatives acting on behalf of the persons who are morally responsible for initiating a pregnancy. Normally, the actions of these persons are performed conscientiously within institutional contexts that provide safeguards for ensuring that any children brought into existence as a result of their actions will be adequately cared for. For instance, in most jurisdictions, fertility services like IVF and artificial insemination by donor (AID) are only provided on the condition that the receiving patient is capable of parenting the child that may result from those services. This condition is legislated in the United Kingdom, for example, through section 14(5) of the *Human Fertilization and Embryology Act* (2008), which states that

> A woman shall not be provided with treatment services unless account has been taken of the welfare of the child who may be born as a result of treatment (including the need of that child for supportive parenting), and any other child that may be affected by the birth.[34]

When services are provided in this type of environment, they are analogous to the actions of the conscientious mine digger in Hanna's example and thus ought to be treated the same with respect to the assignment of responsibility. However, in cases where these safeguards are not in place—in cases, for example, where gametes or fertility services are provided negligently to persons who are clearly incapable of parenting children—it would not be unreasonable to charge the gamete donor or fertility clinician with the responsibility of ensuring that the child is adequately cared for, even if discharging that responsibility required parenting the child themselves. After all, they have voluntarily acted in ways that have foreseeably resulted in the creation of children in need of parental care, and thus satisfy the main conditions for the assignment of moral responsibility.

4.3.2 *The Explanatory Problem*

Even if we can identify a sufficiently narrow constituency of morally responsible causers, a second set of challenges for the causal account relates to the explanatory link it provides between causing a child's existence and acquiring parental obligations. Recall that the causal account is typically defended by way of analogy with everyday cases in which causing a person to be in need generates an obligation to ensure that need is fulfilled. For example, if I hit a pedestrian with my car as a result of my careless driving, then I incur an obligation to ensure that their medical needs are attended to. However, Elizabeth Brake argues that there is a commonly overlooked disanalogy between typical cases of needs creation and the case of causing a child to exist and that this disanalogy speaks against the causal account and in favor of voluntarism. Brake's argument begins with the observation that the obligations generated from typical cases of needs creation tend to be compensatory in nature, where the relevant form of compensation consists in making the victim "whole" again by returning them to their preneed state. On this view, when I hit the pedestrian with my vehicle, the obligation I incur with respect to the victim is to return them to the state they were in *before* I hit them, e.g., by paying for any expenses related to their medical recovery. However, Brake argues that a principle of compensation cannot explain how parental obligations are acquired, for the content of the obligations that attach to the parental role far outstrips compensation for a child's neediness. If a child's neediness consists in the "inability to provide themselves with the necessities of life and to avoid harms and dangers," then compensation for causing a child's neediness would consist in "enabling her to provide herself with necessities, avoid serious harms, and survive independently."[35] However, these compensatory duties—which Brake terms "procreative costs"—are not equivalent to the obligations of parents. As she explains,

> Contemporary parental obligations are extraordinarily weighty in duration and scope ... Our society and legal system assign parents a long support period with responsibilities for more than ensuring mere survival to independence and a minimally decent life ... Bringing children to self-sufficiency—repairing their neediness—does not include the warmth and affection until late adolescence which parents morally owe to their children. Procreative costs could be discharged by providing food, shelter, and basic healthcare, and turning children out the door as soon as they could survive independently...[36]

If sound, Brake's argument appears to put proponents of the causal account on the horns of a dilemma, for it seems like they must either embrace a significantly diminished account of the content of parental obligations or concede that parental obligations are grounded in considerations other than having caused a child's neediness. Brake, of course, believes that causal theorists should take the latter horn and concede that parental obligations are grounded in voluntary acceptance of the parental role, the content of which is determined not by a principle of compensation but rather by social and legal convention. In Brake's view, not only can such an approach support a more robust account of the content of parental obligations, but it is also consistent with their socially constructed nature, or the fact that "what parents owe fluctuates with social change and with the legal institutionalization of parenthood."[37]

How should a causal theorist respond to this argument? One option, of course, is to take the first horn of the dilemma and endorse a significantly diminished account of the content of parental obligations. This approach will likely strike many causal theorists as unappealing, though it might gain plausibility if paired with a principle of supererogation. One might argue, for example, that while it is morally praiseworthy for parents to display attitudes of warmth and affection to their children, these actions are ultimately supererogatory and all that parents are morally required to do for their children by virtue of causing their existence is to repair their neediness by bringing them to self-sufficiency. This line of response may enable causal theorists to account for a diminished account of the content of parental obligations while still maintaining that it is good for parents to display attitudes like warmth and affection.

Of course, if this line of the response still seems unappealing, it is likely because we think that the provision of goods like warmth and affection is *not* in fact supererogatory but rather something that children are entitled to as a matter of justice; and if we think that children are entitled to these goods as a matter of justice, it is likely because we think they *need* them, or because their interest in these goods is sufficiently important to justify the imposition of duties on others to provide them. Thus, a better response to Brake's argument is to challenge her assertion that repairing a child's neediness can be achieved by doing the bare minimum required to bring them to self-sufficiency. Upon reflection, this assertion seems to significantly under-describe children's needs and their inherently relational nature. As Serena Olsaretti has observed,

> to be morally responsible for a child's existence is not only to create a vulnerable creature: it is also to create a vulnerability of a particular kind, due to the fact that virtually all of an infant's needs, and many of a young child's ones, have a strongly relational component.[38]

To bring out the distinctively relational nature of children's needs, Olsaretti contrasts the case of children with the case of an adult who has a physical disability. This adult has certain needs they are unable to satisfy independently (e.g., to move around and to dress themselves) though these needs are not relational in nature and may be adequately satisfied by having one or more persons who they are not in an intimate relationship with assist with certain tasks (or better yet, by having access to technology that would enable them to perform those tasks completely independently). The same is not true of young children. Young children also have needs they are unable to satisfy independently, though from what we know about children's physical and emotional development, these needs are best satisfied in the context of continuous, loving relationships with a small number of adult caregivers.[39] Comparative studies involving institutionalized children reveal the significant impact of a lack of continuous and loving parental care on child development. For example, in a classic series of studies comparing institutionalized children with those raised in families, William Goldfarb found that institutionalized children became ill more frequently, had significant deficits in their learning capacities, became increasingly disinterested in their environment, failed to thrive physically, suffered from insomnia, were often depressed, and eventually developed severe learning disabilities.[40] While there are numerous factors that may have contributed to these deficits, a key factor identified by Goldfarb was the fleeting nature of adult-child interaction among institutionalized children and an attendant lack of "warm, loving contact between a specific parent person and child."[41] Goldfarb's findings have been confirmed in subsequent studies, including the Bucharest Early Intervention Project, the first randomized controlled trial to study the impact of foster care as an intervention for institutionalized children. It found that early placement in high-quality foster care is key to avoiding significant deficits associated with institutional care, including deficits in "cognitive (i.e., IQ) and socio-emotional behaviors (i.e., attachment), brain activity and structure, alterations in reward sensitivity and processing, and a greatly elevated incidence of psychiatric disorders and impairment."[42]

These findings support the view that the need that is created as a result of causing a child to exist is not the need for the provision of basic goods or assistance in the completion of essential tasks (as in the case of an adult with a physical disability), but rather the need for long-term parental care discharged in the context of continuous and loving relationships with a small number of adult caregivers—this, in Velleman's terms, is ultimately what children require in order to cope with their newfound predicament of existence. Once we fully acknowledge the relational nature of children's needs, Brake's objection is neutralized, as it becomes clear that the need that is created as a result of causing a child to exist can only be satisfied by ensuring that a child is adequately cared for in the context of a continuous and loving parent-child relationship.

Even if we take this approach toward resolving Brake's dilemma, however, there are two additional worries about the explanatory power of the causal account that are worth touching upon briefly. First, Brake worries that the causal account cannot explain variability in the content of parental obligations across different societies and time periods. Why, for example, should twenty-first century American procreators owe so much more to their children than their eighteenth-century British counterparts? It might seem unfair that the content of parental obligations should differ so greatly from one society to another or from one time period to another. Second, Brake has also noted that the causal account seems ill-equipped to account for parental obligations outside the context of heterosexual, monogamous, two-parent families. For example, it seems incapable of accounting for parental obligations in the context of matrilineal societies, where maternal uncles are normally taken to be parentally obligated to their sister's children.[43]

It is not clear that either of these worries presents a significant problem for the causal account. With respect to the first worry, variability in the content of parental obligations across different societies and time periods is presumably accounted for by socially determined differences in what is required in order to successfully bring a child to maturity. What is required for a child to flourish in rural England during the eighteenth century is different than what is required for a child to flourish in a twenty-first century American city, and so it is reasonable to think that creating a child in either context would come attached to a substantively different set of obligations. However, it is also possible that people across different societies and time periods have been (and continue to be) mistaken about the nature of children's needs and what constitutes just or appropriate treatment toward them, just as people in different societies and time periods have been (and continue to be) mistaken about what constitutes just or appropriate treatment toward women. The fact that people at different times or in different places have been mistaken about children's needs, such that they have provided children with less than what they require to flourish, does not speak against the explanatory power of the causal account, but simply reveals that its terms have not always been fully observed.

With respect to the second worry, accounting for the obligations that happen to be customary or institutionalized within different social contexts is simply not the role of the causal account of parental obligation, which provides an account of the moral obligations that are generated as a result of causing a child to exist. From the perspective of the causal account, the parental obligations of maternal uncles in matrilineal societies are either explained by their voluntary participation in that social practice (which is in keeping with the pluralistic, indirect version of the causal account), or they simply lack a moral basis insofar as a maternal uncle is not morally responsible for his niece's or nephew's existence and hence need for parental care. The latter possibility speaks less against the explanatory power of the causal account than it does the moral defensibility of matrilineal practices, to the extent they would in fact assign parental obligations to men who have not volunteered for that role and who are not morally responsible for a child's existence.

4.3.3 The Demandingness Problem

If the preceding arguments are sound, then Brake's dilemma is illusory, and the causal account is in fact capable of supporting a more demanding conception of the content of parental obligations, one that includes the duties of warmth and affection that we normally take parents to owe to their children until at least late in their adolescence. However, the weighty and long-term nature of these duties invites a different set of problems. As Porter has emphasized, the causal account "has teeth" in its strict assignment of parental obligations: one does not have to be a willing parent to incur parental obligations to a child, but simply has to stand in a particular causal relationship to that child's existence.[44] I suggested earlier that this is in fact an advantage of the causal account insofar as it allows us to resolve the child abandonment problem, though some may worry that it is excessively demanding by imposing a set of obligations that is disproportionate to the amount of risk taken by engaging in the relevant causal activity—sexual intercourse.[45] The nature of this worry can be helpfully illustrated via an analogy. Like sexual intercourse, driving a vehicle is a desirable and beneficial activity that always presents a risk of an unintended and costly consequence, namely, getting into an accident. However, we normally think that the costs that are incurred as a result of an accident should be proportional to the extent to which the driver is at fault. A driver who injures a pedestrian while driving recklessly or inattentively has acted negligently in a way that leaves them liable to incur the full costs of their actions, whether this takes the form of compensation to the victim or time served in jail. However, a cautious driver who injures a pedestrian in a freak accident is not liable to incur the same costs, despite the fact that the outcome is the same. The fact that they have acted cautiously at least partially absolves them of responsibility for the injury, and this factors into their liability for costs.

The causal account, however, is insensitive to whether or not the relevant causal activities were cautiously or recklessly engaged in. While it assigns full parental obligations to persons who have created a child out of risky or negligent sexual behavior, it may also assign full parental obligations to persons like Alan from *Unwilling Father*—persons who took every reasonable precaution to avoid conception, but whose actions nevertheless resulted in the creation of a child. To the extent that we accept a fault-based view of liability, this insensitivity might seem problematic, for if we are not fully responsible for the outcomes of our actions if we have taken sufficiently rigorous precautions to prevent those outcomes, we might think that we should not have to bear full responsibility for the children we create if we have taken sufficiently rigorous precautions to avoid creating them (e.g., by using multiple reliable forms of birth control).

How should a causal theorist respond to this objection? Two responses are available. The first is to point out that, in practice, the worry about demandingness is largely defused by the various options that persons typically have for permissibly circumventing their parental obligations. For example, a couple who has conceived accidentally might decide to abort the fetus, or, if this is not an acceptable option, they may decide instead to bring the child to term and then delegate caretaking responsibilities to a set of competent and willing adoptive parents. These points go part of the way toward alleviating concerns that participation in even cautious sexual activity leaves one liable for the involuntary acquisition of long-term parental obligations, as there are often viable exit options for those who do not want to undertake them.

However, a second and more direct response is simply to deny the premise that the causal account is excessively demanding. It is true that there are certain cases in which taking reasonable precautions to prevent a particular outcome relieves one of moral responsibility for that outcome in the unlikely event that it comes to pass. However, there are also cases in which we impose strict liability on the outcomes of one's actions, particularly in cases where the actions in question involve significant and inherent risks, including "storing explosives, running nuclear power stations, keeping wild animals, [and] marketing drugs and other dangerous products..."[46] For example, if a tiger in my animal sanctuary escapes from his cage one day and mauls a local resident, I am liable to compensate the victim *even* if I took every necessary precaution to prevent an injury, e.g., by using the strongest and most tamper-proof cages, installing a well-constructed electric fence around the compound, and hiring sufficient staff to patrol the animals and alert local residents in the event of an escape. The reason I am liable in this case is because keeping wild animals is an inherently dangerous activity with a foreseeable chance of resulting in a serious and costly accident. Thus, a condition of permissibly engaging in this activity is the liability to assume those costs in the unlikely but possible event that they arise.

If strict liability is imposed in virtue of the magnitude of the risk associated with a particular activity, then heterosexual intercourse seems like a prime candidate: even when it is undertaken cautiously, it always presents the risk of resulting in the creation of a dependent and vulnerable child with full moral status who may either suffer or thrive depending on whether or not they are adequately parented. Thus, it is reasonable to think that a condition of engaging in that activity should be the obligation to ensure adequate parental care of any child that may result from it, even if every reasonable precaution was taken to avoid that outcome.[47]

Even if it is consistent with analogous cases, however, some may object that a strict liability interpretation of the causal account still seems unfair by assigning the same obligations to cautious procreators as it does to reckless ones—after all, a cautious procreator like Alan has acted morally differently than a reckless procreator who is indifferent to the outcomes of their actions, and we might think that a theory of parental obligation ought to account for that difference. There does seem to be an important moral difference between the conduct of reckless procreators and the conduct of cautious procreators like Alan, though this difference can be accounted for in a way that does not impact the content of their responsibilities. The relevant difference, it seems, involves the extent to which each person is to blame for the outcomes of their actions. Returning to Velleman's analogy, if I accidentally nudge a child into the deep end of a swimming pool, I am less blameworthy than someone who deliberately shoves him in, but our obligations with respect to the child are exactly the same: we are both obligated to jump into the water to fish him out. By the same token, when a person engages cautiously in sexual activity, they may not be blameworthy or subject to moral criticism for the fact that there now exists a vulnerable child in need of parenting, though their obligation to that child is the same as it would be had they acted recklessly: they are still obliged to ensure that the child is adequately parented. In this sense, a person's level of negligence or caution in their procreative activity may affect how we appraise them from the moral point of view, but it need not affect the content of the obligations that activity generates.

4.3.4 *The Asymmetry Problem*

A fourth major challenge for the causal account relates to an asymmetry in the amount of causal influence that men and women typically have over a child's existence. When reflecting on the case of *Unwilling Father*, one reason why it might seem unfair to hold Alan equally parentally liable is that he had an unequal amount of decision-making power over the child's existence: while Alan and Brenda contributed equally to the child's *conception*, it was ultimately Brenda who got to decide whether or not the conceived embryo would be gestated and brought to term. This asymmetry, it seems, is a built-in feature of the causal account. Given the different biological capacities of men and women, and assuming that women have safe access to abortion and robust rights to bodily autonomy, it appears as though women will always have a disproportionate amount of decision-making power over whether or not a child is brought into existence.

How should a causal theorist respond to this asymmetry? This will depend on whether we take the inequality in decision-making power between men and women to be *prima facie* objectionable. If we think it is unfair that men and women have unequal causal influence over a child's existence, yet equal liability to care for the child once it exists, then the causal theorist can respond in one of two ways. First, they can address the inequality in causal influence directly by arguing either (a) that men should be granted equal decision-making power over whether or not a conceived embryo is gestated and brought to term or (b) that they should be given an option to opt out of their parental obligations that is equivalent to a woman's option to choose abortion or adoption, e.g., through being relieved of any responsibility to care for or financially support a child their partner has decided to gestate and bring to term.[48] Alternatively, the causal theorist can also address the implications of the inequality in causal influence by arguing that the inequality in decision-making power should be reflected in the distribution of child-rearing responsibilities. In other words, if women play a greater causal role in the existence of a child, then perhaps they should also incur a greater share of the parental responsibilities.

Neither of these responses are problem-free. Response (a) is clearly a non-starter if we take seriously women's rights to bodily integrity, while the basic idea motivating response (b)—that the option of men to opt out of their parental obligations is equivalent to the option of women to choose abortion or adoption—is based on a conceptual confusion. When a woman chooses to abort a fetus or put a child up for adoption, neither the man nor the woman incurs any child-rearing responsibilities with respect to that child. This is either because the pregnancy is terminated, in which case there are no responsibilities, or because the child is adopted out, in which case the responsibilities are delegated to a willing volunteer. However, when a man chooses to opt out of his parental responsibilities, the same is not necessarily true. In cases where the man opts out, but where his partner continues the pregnancy, she may be fully saddled with the responsibility of parenting the child, a scenario the man is not generally at risk of facing. One might object that if the woman is not prepared to undertake this responsibility, then she can choose to opt out of parenthood as well by terminating her pregnancy or putting the child up for adoption. However, this type of objection only serves to underscore the different realities faced by men and women in procreative decision-making. Whereas men can quite literally walk away from an accidental conception with complete detachment, women are faced with either nine months of pregnancy followed by the relinquishment of their child, or the difficult and often emotionally impactful decision to terminate the pregnancy. To suggest that these are equivalent options—that they constitute a form of "opting out" for women—is to ignore their most important differences.

The second line of response does not fare much better. In addition to the obvious child welfare and gender equality-based reasons against saddling women with a disproportionate share of parental responsibilities, there is also a philosophical reason against doing so: it assumes, as was rejected above, that one's level of moral responsibility for an outcome ought to be proportionate to one's level of causal influence in bringing it about. To the extent that we reject this view in general, we should also be skeptical of its apparent implication that women ought to bear a greater share of parental responsibilities in virtue of their greater biological role in bringing a child into existence.

One might object, of course, that beyond a greater degree of causal influence, women's greater biological role in procreation provides them with a greater degree of *control* over procreative decision-making post-conception, and that one's level of control over the outcomes of one's actions *is* relevant to the assignment of moral responsibility (as per the account of moral responsibility defended in Section 4.3.1). However, it is not clear that an asymmetry in control post-conception justifies a significant asymmetry in the distribution of moral responsibility post-birth, at least in typical unplanned pregnancies resulting from consensual sexual intercourse. After all, the initial act that results in the conception of the child is a joint act over which both parties have an equal degree of control, and the potential consequences of that act—including both an accidental conception and a woman's prerogative to bring an accidental conception to term—are foreseeable to both parties at the time of the act. It is not clear, then, that a man can reasonably object after the fact to the assignment of moral responsibility. To understand the plausibility of this reasoning, consider the following analogy:

Road Trip: Alice and Betty agree to take a road trip together in inclement weather in order to visit their friend during the holidays. They plan to travel in Alice's vehicle, as Betty is unable to drive. Alice is a conscientious driver who always follows traffic laws and drives to the conditions, though the pair encounter a particularly bad snowstorm along the way and, through no fault of her own, Alice loses control of the vehicle, sliding off the highway and causing significant damage to an adjacent property. Betty suggests fleeing the scene in order to spare both travelers the cost of repairing the property, though Alice does not feel comfortable with this given the unique risks it presents to her as the driver and owner of the vehicle. Because Alice is in control of the vehicle, she has the final say on what they will do; as a result, the pair remain on scene to speak with the authorities, where it is determined that Alice, as the driver and owner of the vehicle, is liable for the cost of repairing the property. Alice proposes that Betty share in the cost given that it was incurred as a result of their joint decision to take the road trip together, though Betty refuses on the grounds that Alice could have avoided the costs altogether by driving away from the scene.

Road Trip is an imperfect analogy for an accidental conception, the most important difference being that it would arguably be wrong for Alice and Betty to avoid the costs of their car accident by fleeing the scene, whereas I assume it is not wrong for a couple to avoid the costs of an accidental conception by terminating a pregnancy. Nevertheless, it displays several important features that are analogous to an accidental conception, and which justify treating both cases similarly from a moral point of view. In particular, it is a case in which two agents make a joint decision to engage in an activity that presents a foreseeable risk of incurring certain costs, but in which one party has a disproportionate amount of control over whether those costs are incurred due to their possession of unique capacities—capacities that provide them with a greater degree of decision-making power in the case at hand, but that also come attached to greater burdens and risks. If we think that it would be unreasonable for Betty to refuse to share in the cost of the accident with Alice, then we have a *prima facie* case for extending the same reasoning to the case of an accidental conception and thus rejecting the claim that women's greater degree of control over procreative decision-making post-conception justifies an asymmetry in the distribution of moral responsibility post-birth.

However, while both of the above responses to the causal asymmetry objection ultimately fail, the causal theorist does not in the end need to embrace them, for there is no reason to think that the unequal decision-making power of men and women is objectionable in the first place.

As Millum has noted, unequal treatment of men and women is not objectionable if the inequality is a response to morally relevant differences, and procreation is a case in which sexual differences do seem to make a moral difference. The situation might be described as one in which women have an extra opportunity to exercise a right in virtue of their biological differences.[49] Both men and women have a basic right to bodily integrity that protects their ability to control what happens to and with their bodies, though only women's bodies are capable of conceiving and gestating babies. In the context of procreation, the effect of this difference is that women have an option that men lack: whereas women and men will be equally obligated to any child they jointly cause to exist, women have a bodily-integrity-based right to decide whether a conceived embryo is gestated and brought to term. Moreover, in jurisdictions where this right is legally recognized and publicly known, it is also a factor that men must include in their risk assessment before voluntarily engaging in heterosexual intercourse: that a woman might choose to continue rather than terminate an accidental pregnancy is a reasonably foreseeable consequence of having sex.

Of course, even if there is nothing objectionable about the asymmetrical decision-making power between men and women in procreative contexts, we still might be sensitive to claims of unfairness on the part of unwilling fathers in certain kinds of cases. Don Hubin, for example, worries about cases of accidental pregnancy that involve various forms of deception, e.g., cases in which a woman makes a "lying promise" to terminate any accidental conception that may result from sexual intercourse, but actually plans to proceed with it.[50] It would indeed seem unfair to hold unwilling fathers parentally liable in these types of cases, though the causal account would not in fact support this view insofar as the element of deception undermines the conditions of moral responsibility that ground the assignment of parental obligations in the first place. In this sense, we ought to treat these types of cases similarly to the case of *Sperm Heist* presented above, where one party is solely parentally liable in virtue of their sole moral responsibility for a child's existence.

4.3.5 The Biological Connectedness Problem

A fifth and final challenge applies not to causal accounts in general but rather to the indirect version that I have defended so far in this chapter. The notion that procreators can adequately discharge their obligation to ensure *that* their child is adequately parented by delegating caretaking responsibilities to competent and willing volunteers assumes that children can be adequately parented by persons to whom they are not biologically related and that procreators are not violating any additional moral requirements by alienating themselves from their biological offspring. Initially, both of these seem like relatively safe assumptions. In the first instance, there does not seem to be any reason for thinking that adoptive parents are any less capable than biological parents of providing their children with a supportive upbringing. On the contrary, given the extraordinary lengths that adoptive parents often go to in order to parent children, there may in fact be reason for thinking that they are on average *more* likely than biological parents to provide such an upbringing, given that the latter sometimes find themselves in the parental role unintentionally and hence with less reflection and preparation. In the second instance, there also does not seem to be any reason for thinking that biological parents are necessarily violating any additional moral requirements by alienating themselves from their offspring before or shortly after birth. Provided that children are put up for adoption before developing strong emotional attachments to their biological parents, there does not seem to be any obvious harm associated with "abandoning" them to the care of competent and willing adoptive parents.[51] In cases where the biological parents are themselves incapable of providing a supportive upbringing, this type of "abandonment" may in fact be morally required.

However, some theorists have suggested that biological connectedness matters more than I have acknowledged here, and in a way that impacts on how parental obligations may be appropriately discharged. For example, Bernard Prusak has argued that the "given" relationship between procreators and their progeny is "of tremendous value and importance to [the] child,"[52] as it provides a unique context for the development of deep and unconditional love. On Prusak's view, there is something special about the way that procreators identify with their biological children that leads them to view their task as unconditional, which in turn benefits children by promoting their sense of security and belonging. Prusak concludes on the basis of these claims that "the child's need for emotional support is answered most adequately—and so we could say answered most *responsibly*—by a relationship with his or her biological procreators" and that procreators therefore have a "*prima facie* obligation to develop loving parent-relationships with the children who they presume to bring into being."[53]

Prusak is undoubtedly correct to note that children require access to secure and loving parental relationships in order to flourish. The real question, however, is whether these types of relationships are more likely to obtain between parents and their biological children as opposed to parents and their adoptive children. Prusak does not actually provide any support for this claim, though it is consistent with selected evidence from evolutionary psychology. For instance, some evolutionary psychologists have noted a higher incidence of child maltreatment among step-children as compared to biological children, which they believe can be explained by a parent's evolutionary tendency to invest more heavily in their genetic offspring (a phenomenon known as the "Cinderella Effect").[54] If this explanation is in fact valid and reliably consistent, it might provide a limited basis of support for the *prima facie* obligation to parent one's biological children.

However, it is unclear how much emphasis we should place on these types of explanations, and how much influence we should allow them to have on our theory of parental obligation. For one thing, child maltreatment is a complex phenomenon that is not reducible to a single explanation—indeed, other sources of empirical evidence suggest that a far more significant predictor of child maltreatment is a parent's own history of childhood abuse, a factor that may be distributed evenly among biological and adoptive parents alike.[55] More to the point, however, even if secure and loving relationships *were* more likely to obtain between parents and their biological children, it is not clear how this constitutes an argument against the indirect version of the causal account that I have defended in this chapter. The supposedly greater tendency of biological parents to love and support their children would not imply that delegating parental obligations is impermissible *tout court*, but only that it is impermissible to delegate them to adoptive parents who fail to meet a competency threshold that includes the ability to love and support the child. Prusak's real point, then, is not so much about biology as it is about a child's need to be loved—a need that can be satisfied (or unsatisfied) in the context of biological *and* adoptive parent-child relationships.

However, while Prusak's argument fails to support the notion that parental obligations should be discharged by a child's procreators, it is not the only available account of the value of biological connectedness. An alternative account proposed by Velleman focuses on the special value that is associated with having knowledge of one's biological ancestry, or what he calls one's "family history."[56] According to Velleman, "knowing one's [biological] relatives and especially one's [biological] parents provides a kind of self-knowledge that is of irreplaceable value in the life-task of identity formation."[57] On this view, having a relationship with our biological progenitors is important *not* because they are more likely to love us, but rather because they can provide us with the connection to our ancestral past that we require to form our identities and situate ourselves within a broader narrative. It follows on this account that adoptive parent-child relationships are in an important sense second-best and that it is accordingly "immoral to create children with the intention that they be alienated from their biological relatives."[58]

If biological connectedness matters in the ways that Velleman claims it does, then delegating parental obligations to non-biological adoptive parents may well be problematic. Even if biological parents are not naturally better suited to meet their child's needs than adoptive parents, they may nevertheless be denying their children an important good by alienating them from their family history and the deep self-knowledge it affords. This might again be thought to support a stronger, direct version of the causal account, whereby those who are morally responsible for a child's existence are not only obligated to ensure *that* the child is adequately parented but also obligated to parent the child themselves.

Should we accept this stronger view? Perhaps not necessarily. Even if biological connectedness matters in the way that Velleman claims it does, it does not necessarily follow that children must be *parented* by their biological progenitors. Rather, all that follows is that adopted children must have sufficient access to the knowledge about their biological ancestry that is required for their identity formation and self-conception. For example, it would be consistent with Velleman's account of the value of biological connectedness to permit the delegation of parental obligations on the condition that procreators provide the relevant information to an accessible database (as is now legally required in many jurisdictions for gamete donors)[59] or, if this is insufficient, agree to make themselves available for future contact from their offspring (as in the case of open adoptions). It is also worth noting here that being parented by one's biological progenitors is no guarantee of acquiring the kind of ancestral knowledge that Velleman thinks is so crucial. Just as one can gain knowledge of one's biological roots in the context of an adoptive parent-child relationship, one can lack knowledge of one's biological roots in the context of a biological parent-child relationship, as in cases where one's biological parent is secretive or ignorant about the branches of their own family tree.

Even if biological self-knowledge can be realized in the context of adoptive parent-child relationships, however, we should also be cognizant of the extent to which the value of that knowledge is socially constructed. Both Prusak and Velleman cite the tendency of adopted children to seek out their birth parents as evidence of the deep significance of the biological connection between procreator and progeny, though it is not clear that this is the right conclusion to draw.[60] Rather than underscoring the intrinsic value of biological connectedness, the phenomenon of the "search movement" among adopted children might also be indicative of the higher value that our culture attaches to biological over adoptive relationships, and our tendency to conceive of family history in genetic rather than relational terms. One can imagine that in a world where adoption was more routine, and where adoptive relationships were accorded higher value, there may be less of a desire for adopted children to seek out their biological parents, and perhaps more of a tendency for them to see themselves as part of their adoptive family's lineage. Even in our existing world, where biology happens to take precedence, it is still unclear whether biological self-knowledge is of "irreplaceable value" in the process of identity formation. If this were true, it would imply that children of closed adoptions lead necessarily diminished lives in virtue of being severed from their biological ancestry, though there is no reason to think that this is the case. While some may regret never knowing their biological parents, others develop healthy and robust identities on the basis of non-biological factors, including their own personal history and the new web of relationships they have been adopted into.[61] This should give us pause about the supposed value of biological connectedness, as well as its role within a theory of parental obligation.

Even if these arguments are sound, however, we still might think that there is something morally impoverished about delegating parental obligations to non-biological adoptive parents in certain types of cases. For example, while many would likely agree that delegating parental obligations is permissible or even obligatory in cases where the procreators are unable to adequately care for the child, some may be uncomfortable with the idea that parental obligations may be permissibly delegated in cases where the procreators are able to care for the child, but simply choose not to. I agree that there is a morally relevant distinction between these types of cases, but not one that impacts the latter's permissibility. The elective delegation of parental obligations should only be viewed as impermissible if it involves an instance of wrongdoing, though on the account that I have provided in this chapter, no rights are violated, and no duties are shirked, when a set of procreators decide to delegate custody of their child to a competent and willing recipient. Recall that on the indirect causal account, the obligation that is incurred as a result of causing a child to exist is not the obligation *to* parent the child, but rather the distinct and prior obligation to ensure *that* the child is adequately parented. I have argued that adoptive parents are just as capable as biological parents at parenting children, suggesting that there is no right to be parented by one's procreators, but only a right that one's procreators secure one's parentage. Of course, in line with Velleman's arguments, there may also be a right to know one's genetic origins, but note that this is a different kind of right that imposes different kinds of correlative duties. As I mentioned earlier, one can discharge one's duty to provide information about a child's genetic origins without actually parenting the child.

However, even if the elective transfer of parental obligations is not impermissible, considerations of permissibility do not exhaust the domain of morality, and there may be other bases on which that type of decision can be criticized. For example, while procreators may not be acting unjustly by electively alienating themselves from their biological offspring, they may nevertheless be acting dishonorably by approaching procreation with defective motives,[62] or by failing to appreciate its seriousness and deep moral significance. I do not necessarily agree or disagree with these types of criticisms, but only point out that claiming that an act is morally impoverished is not the same thing as claiming that it is morally impermissible. If there is a general right to do wrong, then we are often permitted to do things that we might have good moral reason not to do.[63] The elective delegation of parental obligations might just be an instance of this more general phenomenon: a decision that possibly reflects poorly on one's character, but does not in fact violate anybody's rights.

4.4 Toward Moral Responsibility for Parental Need

So far, I have defended an indirect version of the causal account of parental obligation, according to which causing a child to exist is sufficient to generate an obligation to ensure *that* the child is adequately parented, which may be discharged either by parenting the child oneself or by delegating caretaking responsibilities to a competent and willing volunteer. This defense has shown that the most prominent objections to the causal account can be adequately responded to; however, it has also revealed that some of the language in which the causal account has traditionally been packaged is inapt and in certain cases misleading. For example, one does not incur the obligation to parent a child simply by causing a child to exist, but rather by satisfying key conditions for the attribution of moral responsibility. Furthermore, one does not incur this obligation in virtue of one's moral responsibility for a child's existence *per se*, but rather one's moral responsibility for the child's need for parental care, a condition that always attends their existence but that is not exclusive to cases of procreation. Thus, I believe it is worthwhile for proponents of the causal account to dispense with the language of causing existence and to reframe their accounts in terms of *moral responsibility for parental need*. This is a more accurate description of what is actually driving the causal account, but the difference is not merely semantic. On the contrary, reframing the causal account in this way directs attention to its morally relevant features, enabling proponents to more adequately respond to objections and to identify previously unrecognized cases in which this account generates obligations.

As an example of the former benefit, consider the *too many parents problem*. Once we acknowledge the distinction between causation and moral responsibility, we can start to see why existing responses to addressing the too many parents problem are misguided. For example, Austin, Nelson, and Porter have all attempted to address this problem by proposing modified conceptions of causation that count a smaller number of people as causes: Austin and Nelson suggest that we should focus on "primary and proximate" causes, while Porter suggests that we should embrace so-called INUS causation, according to which something counts as a cause of an outcome if it is an insufficient but necessary part of the unnecessary but sufficient condition of that outcome.[64] However, these strategies will only succeed if we assume—implausibly—that being a cause of a particular outcome is sufficient for bearing moral responsibility for it. On the contrary, refocusing the account on moral responsibility directs our attention to a more relevant set of questions, namely, what conditions must obtain for a person to be morally responsible for a child's existence and hence the need for parental care.

As an example of the latter benefit, consider a non-procreative case in which being responsible for a child's need for parental care also generates parental obligations:

> *Drunk Driver*: John is driving home one night after he has had far too much to drink. He runs a red light and crashes into another car containing a family of recent immigrants, where the two parents are killed and the two children survive.

Just like a procreator who is morally responsible for a child's existence, John is morally responsible for the children's need for parental care, and thus it is reasonable to think that he also incurs an inalienable obligation to ensure that need is fulfilled for the duration of their childhood. Of course, there may be social or legal institutions in place that ensure this need is fulfilled or that facilitate the performance of John's obligation, e.g., by extracting child support payments through the garnishment of his wages. However, in the absence of these obligations, John would still retain a weighty obligation to ensure that the children are adequately parented, suggesting that the scope of application of a moral responsibility for parental need account of parental obligation is broader than the case of procreation.

4.5 Parenthood Unpackaged

Having provided an account of the moral relationship between procreation and parental rights, as well as the moral relationship between procreation and parental obligation, we are now in a position to draw some general conclusions. At the beginning of Chapter 3, I introduced what David Archard calls the *parental package* thesis, which holds that the kinds of considerations that serve to ground parental rights are the same kinds of considerations that serve to ground parental obligations.[65] On this view, if causing a child to exist through procreation generates a special right to parent that child, then it also generates a special obligation to parent that child as well and vice versa. We can now see that this common view is in fact false and that parental rights and obligations have different moral foundations. While the presumptive right to parent one's biological child is rooted in considerations of autonomy and the more general right to continue one's justice-respecting projects, the presumptive obligation to parent one's biological child is typically rooted in one's moral responsibility for that child's need for long-term parental care.

As an illustration of how these two features come apart, consider an important test case. According to the National Conference of State Legislatures (NCSL), there are normally between 17,000 and 32,000 rape-related pregnancies in the United States each year.[66] States themselves are split in terms of how they legally assign parental rights in cases where pregnancies are carried to term, with roughly half of states allowing for the outright termination of perpetrators' rights, and half allowing for significant restrictions on custody and visitation—all of which is usually conditional on a successful conviction.[67] As legal parental obligations are typically assigned on the basis of genetic connectedness or voluntary acknowledgment of parentage, it will usually be the case that perpetrators also incur legal parental obligations to children conceived from rape.

How should we appraise these legal conventions from a moral point of view? Consider first the issue of parental rights. Despite their legal acknowledgment in some states, it seems implausible that a rapist would have any moral grounds for claiming a right to parent a child conceived out of rape. One explanation for this view is suggested by the type of justification state legislatures provide for existing legislation aimed at terminating or limiting the parental rights of rapists, which typically focuses on the best interests of the child. On this view, rapists do not have a presumptive right to parent their victim's child because their history of sexual violence renders them unfit to parent children generally. This is a valid argument and is perhaps sufficient to support the forfeiture or non-assignment of parental rights in most (if not all) cases of rape. However, there is also an additional argument against assigning parental rights to rapists based on the parental project view, which can be gleaned from structurally analogous cases. Consider, for example, one such case:

> *Embryo Creator*: Kevin breaks into a gamete storage facility and steals an ovum that Janet had frozen for a future procreative project. He then combines it with his own sperm *in-vitro*, and has his partner gestate the resulting embryo for nine months, leading to the birth of a healthy baby who he now intends to parent.

In this instance, even though Kevin has contributed some of the genetic material from which the child developed, his parental rights claim to the child seems illegitimate for the same reasons that Dana's parental rights claim was illegitimate in the case of *Gamete Thief* from Chapter 3, namely, because the child was created from the use of gametes to which he did not have legitimate access. Thus, the parental project he has initiated with respect to the child is not justice-respecting in the relevant sense. Even if it is not primary, this point at least serves as a supplementary consideration weighing against the assignment of parental rights to rapists. By impregnating a victim as a result of rape, the rapist engages in the same type of wrongdoing as Kevin in *Embryo Creator*, in addition to the wrongdoing associated with the assault itself: he forces the creation of an embryo with reproductive gametes to which he does not have legitimate access. In this sense, we should not think of a rapist as someone who has presumptive parental rights that are subject to being overridden by considerations of child welfare; rather, we should think of him as someone who does not have any presumptive parental rights to begin with, due to the manner in which his genetically related child was conceived.

Consider next the issue of parental obligations. While both Kevin and the rapist do not appear to have any legitimate right to parent the child they played a role in creating, their moral responsibility for that child's existence (and hence the need for parental care) suggests that they are still liable to ensure the child is adequately parented. However, they do not stand in the appropriate rights position with respect to the child, and they cannot discharge this obligation by parenting the child themselves; rather, they must see to it the child is adequately parented by someone else, e.g., by providing child support payments to the child's mother. Of course, the victim of rape can plausibly be understood to be in the opposite position. To the extent that she cannot be held morally responsible for the child's existence, she does not incur an obligation to parent the child or to ensure that the child is adequately parented. However, insofar as she has not engaged in any wrongdoing with respect to the conception of the child, any parental intentions that she forms with respect to the child ought to be respected.

Recognizing the distinction between the basis of parental rights and obligations allows us to draw two general conclusions. First, in cases where there are multiple conflicting claims over who gets to parent a child, parental rights should be assigned primarily on the basis of legitimate parental intentions. This suggests that the status quo on surrogacy dispute resolution is currently backward, as priority is currently given to the woman who gestated the baby, regardless of whether that baby is genetically related to her.[68] Second, in cases where there is disagreement between different parties about who is responsible for parenting a particular child, parental obligations should be assigned in the first instance to whoever is morally responsible for the child's need for parental care. This suggests that the legal assignment of parental obligations is normally correct, but that the explanation supporting it is wrong. Parental obligations do not derive from the contribution of the genetic material from which a child developed, but rather from one's moral responsibility for that child's need for parental care. While these factors often overlap, failing to distinguish between them can lead one to unjustified and counterintuitive conclusions, such as the assignment of parental obligations to victims of sexual assault, gamete theft, or deceptive sexual practices, and the assignment of parental rights to rapists, gamete thieves, and rogue gestational surrogates.

Notes

1 David Benatar. "The Unbearable Lightness of Bringing into Being," *Journal of Applied Philosophy*, 16(2) (1999), pp. 173–180; Melissa Moschella. "Rethinking the Moral Permissibility of Gamete Donation," *Theoretical Medicine and Bioethics*, 35(6) (2014), pp. 421–440 2014; and Rivka Weinberg, "The Moral Complexity of Sperm Donation," 22(3) (2008), pp. 166–178 and *The Risk of a Lifetime: How, When, and Why Procreation May Be Permissible* (New York: Oxford University Press, 2015), esp. chapter 2.

2 Elizabeth Brake. "Fatherhood and Child Support: Do Men Have a Right to Choose?" *Journal of Applied Philosophy*, 22(1) (2005), pp. 55–73, and "Willing Parents: A Voluntarist Account of Parental Role Obligations" in David Archard and David Benatar (eds.) *Procreation and Parenthood: The Ethics of Bearing and Rearing Children* (Oxford: Oxford University Press, 2010); Onora O'Neill. "Begetting, Bearing, and Rearing" in Onora O'Neill and William Ruddick (eds.) *Having Children: Philosophical and Legal Reflections on Parenthood* (New York: Oxford University Press, 1979); and Judith Jarvis Thomson. "A Defense of Abortion," *Philosophy and Public Affairs*, 1(1) (1971), pp. 47–66.

3 David Archard. "The Obligations and Responsibilities of Parenthood" in Archard and Benatar, 2010; Michael W. Austin. *Conceptions of Parenthood: Ethics and the Family* (Burlington: Ashgate, 2007), chapter 3; Jeffrey Blustein. "Procreation and Parental Responsibility," *Journal of Social Philosophy*, 28(2) (1997), pp. 79–86; Daniel Callahan. "Bioethics and Fatherhood," *Utah Law Review*, 3 (1992), pp. 735–746; Giuliana Fuscaldo. "Genetic Ties: Are They Morally Binding?" *Bioethics*, 20(2) (2006), pp. 64–76; James Lindemann Nelson. "Parental Obligations and Surrogacy Contracts: A Causal Perspective," *Public Affairs Quarterly*, 5(1) (1991), pp. 49–61; Lindsey Porter. "Adoption Is Not Abortion-Lite," *Journal of Applied Philosophy*, 29(1) (2012), pp. 63–76, and "Why and How to Prefer a Causal Account of Parenthood," *Journal of Social Philosophy*, 45(2) (2014), pp. 182–202; Bernard Prusak. "The Cost of Procreation," *Journal of Social Philosophy*, 42(1) (2011), pp. 61–75; and J. David Velleman. "Persons in Prospect," *Philosophy and Public Affairs* 36(3) (2006), pp. 221–288, esp. 245–266.

4 Teresa Baron. "A Lost Cause? Fundamental Problems for Causal Theories of Parenthood," *Bioethics*, 34(7) (2020), pp. 664–670.

5 This typology is presented by Bayne and Kolers, "Towards a Pluralist Account of Parenthood," 221–222.

6 In three-parent IVF, the nucleus of a donor egg is inserted into the cytoplasm of a recipient egg from which the nucleus has been removed due to mitochondrial abnormalities. The hybrid egg is then fertilized with a man's sperm, thereby giving the resultant child three genetic parents.

7 See Fuscaldo's discussion of *Magill v. Magill*, in "Genetic Ties," 65.

8 See, for example, Benatar, "The Unbearable Lightness," and Weinberg, "The Moral Complexity of Sperm Donation."

9 Joseph Millum. "How Do We Acquire Parental Responsibilities?" *Social Theory and Practice*, 34(1) (2008), pp. 71–93, 72.

10 Weinberg, "The Moral Complexity of Sperm Donation."

11 Weinberg, "The Moral Complexity of Sperm Donation," 170.

12 Michael Hardimon. "Role Obligations," *Journal of Philosophy*, 91(7) (1994), pp. 333–363, 343. Note that Hardimon is merely reporting this principle and does not himself endorse it. For a qualified defense of the volunteer principle with respect to special obligations generally, see Diane Jeske. "Families, Friends, and Special Obligations," *Canadian Journal of Philosophy*, 28(4) (1998), pp. 527–556.

13 See Hardimon, "Role Obligations" and Christina Hoff Sommers. "Filial Morality," *The Journal of Philosophy*, 83(8) (1986), pp. 439–456.

14 As Millum points out in the context of his *conventional acts* account of parental obligation, the particular actions that may signal the assumption of a parental role can be socially or culturally dependent. See "How Do We Acquire Parental Responsibilities?".

15 Brake, "Willing Parents," 172–173.

16 See, for example, Brake, "Father and Child Support," and Stephen D. Hales. "Abortion and Father's Rights" in James M Humber and Robert F. Almeder (eds.) *Reproduction, Technology, and Rights* (Totowa: Humana Press, 1996).

17 See, for example, Austin, *Conceptions of Parenthood*, 55, and Porter, "Adoption is not Abortion-Lite," 66.

18 Thomson, "A Defence of Abortion," 65.

19 Brake, "Willing Parents," 174.

20 Velleman, "Persons in Prospect," 251.

21 There is a possible objection here that being the cause of a person's existence is distinct from being the cause of their neediness and that while a procreator may be responsible for the existence of a child, they are not responsible for that child's need for parental care. This may be true in a strict sense, though given the facts that (a) a child's existence and neediness always go together (one cannot create a *non*-needy child) and (b) their neediness is foreseeable, this seems to be a distinction without a difference. For a defense of this type of argument, see Brake, "Fatherhood and Child Support," 60, and Millum, "How Do We Acquire Parental Responsibilities?" 76–77. For a comprehensive critique, see Jeff McMahan. *The Ethics of Killing: Problems at the Margins of Life* (Oxford: Oxford University Press, 2002), 364–372.

22 This distinction between transferring and delegating parental obligations is presented by Reuven Brandt in "The Transfer and Delegation of Responsibilities for Genetic Offspring in Gamete Donation," *Journal of Applied Philosophy*, 34(5) (2017), pp. 665–678. For an example of an indirect causal account that endorses the *transfer view*, see Archard, "Th Obligations and Responsibilities of Parenthood," esp. 105 and 116–118. For an example of an indirect causal account that endorses the *delegation view*, see Porter, "Adoption Is Not Abortion-Lite," esp. 70–73, and "Why and How to Prefer a Causal Account of Parenthood," esp. 191–196.

23 I will argue for this claim below in Section 4.3.5.

24 David Lewis. "Causation," *Journal of Philosophy*, 70(17) (1973), pp. 556–567.

25 See Archard, "The Obligations and Responsibilities of Parenthood," 112; Austin, *Conceptions of Parenthood*, 39; Brake, "Willing Parents," 158; Blustein, "Procreation and Parental Responsibility," 82–83; Fuscaldo, "Genetic Ties," 67; and Nelson, "Parental Obligations and Surrogacy Contracts," 53–54.
26 Blustein, "Procreation and Parental Responsibility," 82.
27 Austin, *Conceptions of Parenthood*, 39. See also Nelson, "Parental Obligations and Surrogacy Contracts," 54.
28 It is worth noting, however, that not all theorists agree that causation is even necessary for moral responsibility. For an argument against this view, see Carolina Sartorio. "How to Be Responsible for Something Without Causing It," *Philosophical Perspectives*, 18 (2004), pp. 315–336.
29 See, for example, Fernando Rudy-Hiller. "The Epistemic Condition for Moral Responsibility," *Stanford Encyclopedia of Philosophy* (Winter 2022 Edition), Edward N. Zalta and Uri Nodelman (eds.), available from https://plato.stanford.edu/archives/win2022/entries/moral-responsibility-epistemic.
30 Note that in instances where this was not the case, we would not hesitate to assign moral responsibility. Imagine, for example, that Albert intentionally sabotages Beth's high school romance by starting a rumour of infidelity about her boyfriend Christopher, which eventually leads to an erosion of trust and a breakdown in the relationship. Albert's action was not proximate to the eventual outcome—the rumour required subsequent action by numerous intermediaries in the high school gossip chain to reach Beth, after which it required further action on the part of Beth and Christopher to result in a breakup—though insofar as this sequence of events was reasonably foreseeable at the time of Albert's action, it is still appropriate to hold him morally responsible for the outcome.
31 It might be objected here that if voluntary participation in the relevant causal sequence is necessary for incurring parental obligations, then the causal account simply collapses into a form of voluntarism. However, this objection is based on a conceptual confusion. What is being volunteered for in the present context is not the role of acting as a child's parent, but rather an act that can foreseeably result in the creation of a child who needs to be parented. One can involuntarily incur obligations as a result of voluntarily engaging in an act that leaves one liable to incur them (think, for example, of the obligation to compensate for damages as a result of causing a car accident).
32 See, for example, Victor Durà-Vilà. "Parental Obligation, Adoption, and Abortion: Critique of Porter and a Nozickean Alternative," *Journal of Value Inquiry*, 47(1–2) (2013), pp. 29–47, and Jason Hanna. "Causal Parenthood and the Ethics of Gamete Donation," *Bioethics*, 33(2) (2019), pp. 267–273, esp. 269.
33 Hanna, "Causal Parenthood and the Ethics of Gamete Donation," 271.
34 United Kingdom. *Human Fertilization and Embryology Act 2008*, s. 14(5).
35 Brake, "Willing Parents," 160.
36 Brake, "Willing Parents," 160.
37 Brake, "Willing Parents," 160.
38 Serena Olsaretti. "Liberal Equality and the Moral Status of Parent-Child Relationships," in David Sobel, Peter Vallentyne, and Steven Wall (eds.) *Oxford Studies in Political Philosophy, Volume 3* (Oxford: Oxford University Press, 2017), 73–74.
39 See, for example, Sue Gerhardt. *Why Love Matters: How Affection Shapes a Baby's Brain* (New York: Routledge, 2004). S. Matthew Liao also provides a comprehensive overview of the developmental importance of parental love for children in *The Right to Be Loved* (Oxford: Oxford University Press, 2015), chapter 3.

40 This summary is drawn from S. Matthew Liao's comprehensive overview in *The Right to Be Loved*, chapter 3. See also William Goldfarb. "Effects of Psychological Deprivation in Infancy and Subsequent Stimulation," *American Journal of Psychology*, 102(1) (1945), pp. 18–33.
41 Goldfarb, "Effects of Psychological Deprivation," 18.
42 See Bucharest Early Intervention Project. "About the Bucharest Early Intervention Project," accessed July 2023 from https://www.bucharestearlyinterventionproject.org/.
43 Brake, "Willing Parents," 166–167.
44 Porter, "Why and How to Prefer a Causal Account of Parenthood," 183.
45 For expressions of this worry, see Brake, "Fatherhood and Child Support," 60, and Millum, "How Do We Acquire Parental Responsibilities?" 72–73.
46 Tony Honoré. *Responsibility and Fault* (Oxford: Hart Publishing, 1999), 23.
47 Strict liability interpretations of the causal account have been explored in Archard, "The Obligations and Responsibilities of Parenthood," 115–116, and Blustein, "Procreation and Parental Responsibility," 84–85. Rivka Weinberg has also embraced a strict liability approach in the context of her "Haz-Mat" theory of parental obligation. See Weinberg, "The Moral Complexity of Sperm Donation," and *The Risk of a Lifetime*, chapter 2.
48 Versions of the latter view are defended in Brake, "Fatherhood and Child Support," and Hales, "Abortion and Father's Rights."
49 Millum, "How Do We Acquire Parental Responsibilities?" 90–93.
50 Don Hubin. "Procreators' Duties: Sexual Asymmetries" in Leslie Francis (ed.) *The Oxford Handbook of Reproductive Ethics* (Oxford: Oxford University Press, 2017), esp. 316–317.
51 As the proviso in this sentence suggests, things are different if a child is put up for adoption *after* developing strong emotional attachments to their biological parents, as this kind of detachment is likely to harm the child emotionally and compromise their interest in continuity of care. Of course, doing so may still be justified under certain circumstances, e.g., if the biological parents become unable to care for the child.
52 Bernard Prusak. *Parental Obligations and Bioethics: The Duties of a Creator* (New York: Routledge, 2013), 19.
53 Prusak, *Parental Obligations and Bioethics*, 36.
54 Martin Daly and Margo Wilson. *The Truth about Cinderella: A Darwinian View of Parental Love* (New Haven, CT: Yale University Press, 1999).
55 See, for example, Angela Moreland Begle, Jean E. Dumas, and Rochelle F. Hanson. "Predicting Child Abuse Potential: An Empirical Investigation of Two Theoretical Frameworks," *Journal of Clinical Child & Adolescent Psychology*, 39(2) (2010), pp. 208–219.
56 See J. David Velleman. "Family History," *Philosophical Papers*, 34(3) (2005), pp. 357–378.
57 David Velleman. "Family History," 357.
58 David Velleman. "Family History," 357.
59 For example, Austria, Finland, New Zealand, Norway, Sweden, Switzerland, and the United Kingdom all legally require gamete donors to be identifiable to their genetic offspring. For a discussion of this requirement and the possible ethical considerations that underlie it, see Inmaculada de Melo Martin. "The Ethics of Anonymous Gamete Donation: Is There a Right to Know One's Genetic Origins?" *Hasting Center Report*, 44(2) (2014), pp. 28–35.

60 See Prusak, *Procreative Obligations and Bioethics*, 15; and Velleman, "Family History," 359–360.
61 For a more extended critique of Velleman's arguments, see Sally Haslanger. "Family, Ancestry and Self: What is the Moral Significance of Biological Ties?" in *Resisting Reality: Social Construction and Social Critique* (Oxford: Oxford University Press, 2012).
62 For an exploration of the role that motives play in the morality of procreation, see David Wasserman. "Pro-Natalism" in David Benatar and David Wasserman (eds.) *Debating Procreation: Is it Wrong to Reproduce?* (Oxford: Oxford University Press, 2015), esp. chapter 8.
63 Jeremy Waldron. "A Right to Do Wrong," *Ethics*, 92(1) (1981), pp. 21–39.
64 See Austin, *Conceptions of Parenthood*, 39; Nelson, "Parental Obligations and Surrogacy Contracts," 54; and Porter, "Adoption is Not Abortion-Lite," 68–69.
65 Archard, "The Obligations and Responsibilities of Parenthood,"106.
66 National Conference of State Legislatures. "Parental Rights and Sexual Assault," available from https://www.ncsl.org/human-services/parental-rights-and-sexual-assault.
67 The fact that the termination or restriction of parental rights is usually conditional on a successful conviction suggests a disturbingly low incidence of enforcement. According to data gathered from the United States Department of Justice, more than 68% of sexual assaults go unreported, and of the remaining 32% that are reported, less than 2% result in successful convictions. For a summary overview, see the Rape, Abuse, and Incest National Network (RAINN). "Reporting Rates," available from https://rainn.org/get-information/statistics/reporting-rates.
68 See, for example, s. 33(1) of the United Kingdom's *Human Fertilization and Embryology Act* (2008), which states that "The woman who is carrying or has carried a child as a result of the placing in her of an embryo or of sperm and eggs, and no other woman, is to be treated as the mother of a child."

5 Distributing the Cost of Procreation

5.1 Introduction

Anyone who has undertaken the project of creating and raising a child will be all too familiar with the fact that kids are expensive. A 2017 report by the United States Department of Agriculture (USDA) found that the average cost of raising a child in that country from birth until the age of 17 is in excess of $233,610, which averages out to approximately $13,742 per year, $1145 per month, and $38 per day.[1] For the typical two-parent family, this expenditure accounts for around 16% of annual household income; for single-parent families, it exceeds 37%.[2] And these, of course, are only the financial costs. In addition to a significant financial investment, prospective parents can also expect to invest an enormous amount of time and energy in the raising of their children and to incur innumerable opportunity costs along the way. They can expect to routinely forgo their own interests in favor of the interests of their children and to live their lives with less flexibility than non-parents in their society who are in similar financial circumstances.

From the perspective of justice, an important question concerns how the significant costs of procreation and parenthood should be distributed throughout society: should parents be responsible for paying the full cost of their child-rearing projects, or should child-rearing costs be shared in some way among parents and non-parents alike? To the extent that philosophers have considered this question, two sets of views have emerged, which, following Serena Olsaretti, we can label *anti-cost-sharing* and *pro-cost-sharing* views (or anti-sharing and pro-sharing views, for short).[3] According to anti-sharing views, the cost of creating and raising children should be borne primarily, if not entirely, by parents themselves. This set of views counts among its defenders responsibility-sensitive egalitarians, who see child-rearing as a choice for which individuals ought to be held responsible.[4] Assuming an initially just distribution of resources, these theorists reason that it would be unfair to diminish the resource share of

DOI: 10.4324/9781003164104-6

non-parents for the sake of financing the expensive choices of parents—after all, it is parents who brought it about that there are children with costly needs to be met, so the responsibility for meeting those needs ought to fall primarily on them.

Proponents of pro-sharing views see things differently. While it is true that the costs of child-rearing arise out of the voluntary[5] choices of parents, focusing too narrowly on the origins of those costs distracts attention from the various ways in which non-parents benefit from others' child-rearing labor. In societies with redistributive social welfare systems, for example, non-parents benefit from the existence of a younger generation that can maintain the economic institutions they will depend on in old age.[6] They may also be considered the past beneficiaries of the child-rearing labor of their *own* parents, to the extent that being brought into existence and raised into an adult can be considered a benefit.[7] Proponents of pro-sharing views insist that because non-parents stand to benefit from others' child-rearing labor in either or both of these ways, they have a justice-based obligation to share in the cost of that labor by contributing to some form of cost-sharing scheme.

The position we take on the question of how to distribute the costs of procreation and parenthood will have important ripple effects throughout a broader theory of procreative justice. I argued in previous chapters that the right to procreate is constrained by a prior obligation to ensure that any child created as a result of its exercise will have a decent start in life and that being morally responsible for a child's existence (and hence need for parental care) always generates an obligation to ensure that the child is adequately parented, either by parenting the child oneself or by delegating caretaking responsibilities to a capable and willing volunteer. The extent to which prospective procreators are capable of discharging these obligations will depend in a significant way on how the society they live in distributes their associated costs. For example, in societies that externalize costs through the provision of benefits like child subsidies, paid parental leave, or universal childcare, morally permissible procreation will be available to more prospective parents, whereas in societies that internalize costs by holding parents primarily liable, it will be available to far less. This highlights the important though sometimes underappreciated fact that the moral permissibility of procreation turns not only on features of the personal circumstances or capacities of procreators themselves but also on features of the broader social context in which they make their procreative decisions.

In this chapter, I address one feature of this social-contextual question by considering how the costs of procreation and parenthood ought to be distributed throughout society. I ultimately defend a forward-looking version of the pro-sharing view, according to which non-parents are obligated to share in the costs of child-rearing in virtue of their status as the future beneficiaries of parents' child-rearing labor. However, I also argue that the tenability of this view is conditional on two further considerations. First, it is conditional on the success of independent arguments for redistributive social welfare institutions that translate child-rearing labor into benefits for others; and second, it is conditional on others actually benefitting from those institutions, an empirical consideration that may vary significantly according to both the rate of procreation and how we conceive of our broader demands of global and intergenerational justice. After demonstrating why a forward-looking pro-sharing view is preferable to the anti-sharing view, I attempt to show how both of these conditions can be met, and how non-parents are therefore obligated to share (to a certain extent) the cost of parents' child-rearing projects.

The chapter proceeds in four parts. In Section 5.2, I consider the most prominent version of the anti-sharing view, namely, the view that child-rearing is a choice whose costs ought to be internalized by the individuals who decide to pursue it. In Section 5.3, I consider the most prominent pro-sharing response to this view, namely, the view that children are public goods whose costs ought to be externalized among the third parties who stand to benefit from them. In Sections 5.4 and 5.5, I offer a qualified defense of one variant of this view—the *socialized goods account*—by showing how it can meet the two aforementioned conditions.

5.2 The Responsibility Argument

Consider the following case:

> *Desert Island*: A small group of shipwreck survivors have washed up on the shore of a desert island and are forced to establish a new society together. Recognizing that no one has any special entitlement to the island's resources, they collectively decide to distribute the resources equally, so that each islander has an equal bundle that they can use in ways consistent with their own preferences and ambitions. Since each islander has different tastes and ambitions, each uses their resources for different purposes, which in turn has different effects on the overall distribution: entrepreneurial islanders invest their resources in risky commercial ventures in the hopes of earning significant returns; prudent islanders use their resources sparingly, saving rations that can sustain them in the future; and leisure-prone islanders live off their resources freely, as they spend their days swimming and basking in the sun. Because each islander starts with an equal bundle of resources, and only loses or gains resources as a result of voluntary exchanges, it is assumed that the resulting distribution is just.[8]

Now suppose that one pair of islanders decide to create a child in order to fulfill their personal ambition of starting a family. What effect should this decision have on the overall distribution of resources? On the one hand, given that the child is a person, we might think of her as a late-coming islander who is also entitled to an equal share of the island's resources. On this view, the resource share of each islander ought to be diminished to accommodate the new arrival, just like it would be if there was one additional shipwreck survivor at the time of the initial allocation. On the other hand, given that the child was created deliberately to service her parents' ambition of starting a family, we might think of her as the kind of voluntary decision for which individuals ought to be held responsible. On this view, the cost of caring for her should not be distributed equally among the other islanders, but should rather be internalized by the particular individuals who are responsible for her existence. Eric Rakowski defends a version of this latter view, claiming,

> ...babies are not brought by storks whose whims are beyond our control. Specific individuals are responsible for their existence. It is therefore unjust to declare ... that because two people decide to have a child, or through carelessness find themselves with one, *everyone* is required to share their resources with the new arrival, and to the same extent as its parents. With what right can two people force all the rest, through deliberate behavior rather than bad brute luck, to settle for less than their fair shares after resources have been divided justly? If the cultivation of expensive tastes, or silly gambles, or any other intentional action cannot give rise to redistributive claims, how can procreation?[9]

Rakowski's view is rooted in a particular brand of responsibility-sensitive (or "luck") egalitarianism, one that holds individuals responsible for the costs of their choices when made against a fair background distribution of resources. On this type of view, individuals have legitimate claims to compensation for inequalities arising from bad "brute luck" (or luck that befalls them regardless of their choices), but they do not have legitimate claims to compensation for inequalities arising from bad "option luck" (or luck in the way their calculated gambles turn out), nor for those arising from the satisfaction of expensive tastes.[10] In this sense, just as it would be unfair to force the community of islanders to share the cost of a failed commercial venture, or to subsidize a lazy neighbor's beach-going lifestyle, so too would it be unfair to force them to share the cost of others' procreative decisions.

It is important to note here that endorsing the general idea that parents should be responsible for paying the cost of their procreative decisions does not necessarily commit one to a specific idea about *which* costs they are responsible for paying. As Olsaretti has pointed out, any theory of responsibility must include two separable elements: an *attribution principle*, which explains *who* is responsible for the consequences of a particular action or choice; and a *principle of stakes*, which explains *which* particular consequences they are responsible for.[11] Proponents of the responsibility argument may agree on the principle of attribution, or the idea that responsibility falls on the persons who choose to bring a child into existence, but disagree on the principle of stakes, or the particular child-rearing costs those persons should be responsible for paying.[12] In what follows, I assume that if parents are responsible for paying the costs of their child-rearing projects, they are only responsible for the costs associated with providing their children with what they are justly entitled to. These will include what Olsaretti calls the "costs of care," or the time, money, and energy that are required to adequately care for children during their childhood, as well as the "costs of added adult members," or the costs associated with providing children with what they are justly due as adults.[13]

With this caveat in mind, is the responsibility argument sound? On the one hand, it seems to capture a strong intuition that many people will share about choice and personal responsibility. If we do not think that individuals have legitimate claims to compensation for the other expensive choices—say, the choice to climb Mount Everest—then it is not clear why they should have legitimate claims to compensation for their procreative choices. After all, the cost of each choice is known (or at least knowable) beforehand, and individuals are free to forgo that choice if they consider its costs to be prohibitive. On the other hand, however, the responsibility argument might also seem counterintuitive, for it appears to entail a radical revision in the types of costs that parents are liable for paying. As Patrick Tomlin notes, "Is it really true that schools and children's hospitals should ideally be entirely parentally funded? Or that state funded parental leave (against a just background distribution) is an affront to justice?"[14]

While many people will likely share Tomlin's skepticism, it is important to note that the bare fact that the responsibility argument has radical implications does not necessarily constitute an objection to that argument. A proponent could easily respond that the responsibility argument is most plausibly pitched at the level of ideal theory: it purports to explain what our child-rearing responsibilities would be if, contrary to fact, resources were distributed justly, and in such a society it is not clear that the implications of the responsibility argument would be unacceptable. However, while Tomlin's objection misses the mark, there are other possible objections that take the responsibility argument on its own terms. For example, G.A. Cohen has notably argued that while responsibility sensitive egalitarianism prohibits compensation for expensive tastes that have been deliberately cultivated, it does not necessarily prohibit compensation for expensive tastes that a person cannot be held responsible for developing.[15] If Cohen is right, then one way to resist the implications of the responsibility argument is to claim that the desire to have children falls into the second category, either because the desire to procreate is a biologically driven instinct or because it is the product of social norms and pressures for which individuals cannot be held responsible.[16] Alternatively, one might also distinguish between child-rearing and other expensive projects on perfectionist grounds, or in virtue of the value that is created or involved in their pursuit. If we distinguish between different types of projects in terms of their relative value, it is not clear that creating and raising children should be treated on par with drinking champagne or eating plover eggs.

Both of these objections are tenable, though neither is problem-free. While one might be able to argue that there is a range of external factors that contribute toward the desire to have children, the significance of these factors is undermined by the fact that the vast majority of persons who choose to have children identify with and reflectively endorse their choice. Moreover, while there seems to be an obvious difference between the choice to have children and the choice to pursue a luxury lifestyle, it is harder to distinguish between child-rearing and some of the other expensive projects that a responsibility principle is wont to cover, particularly for those who are committed to a stance of neutrality between competing conceptions of the good life.[17] Just as child-rearing is costly, time-consuming, and central to the identity and self-conception of prospective parents, so too is the choice to climb Mount Everest for many mountaineers. One might be able to draw a principled distinction between these two types of projects, but not without taking a controversial ethical stand that will be open to criticisms by proponents of liberal neutrality, who believe the state ought to remain neutral between competing conceptions of the good life. There is, however, a more promising way of distinguishing between child-rearing and other expensive projects that appeals to the benefits they will create for others. This is the starting point for the most prominent version of the pro-sharing view.

5.3 The Public Goods Argument

In framing his initial argument against cost sharing, Rakowski is adopting a single-generation perspective, assuming that the non-parents of a particular generation will be made unjustifiably worse off by having to share in the cost of child-rearing due to a decrease in their current overall resources. This may be true in the short term; yet it underplays the fact that, in the longer term, many of those children will grow up to become productive members of society, and the fruits of their labor may be used to support members of the previous generation in old age. Once we adopt this multi-generational perspective, we see that child-rearing is importantly distinct from the other expensive projects that individuals might pursue. Few, if any, people stand to benefit from another person's project to climb Mount Everest. Yet most, if not all, people stand to benefit from the existence of a younger generation that can maintain the social and economic institutions they will depend on in old age.

A simpler way of putting the point is to say that decisions to bear and rear children tend to produce positive externalities that benefit parents *and* non-parents alike. Nancy Folbre describes this effect in the following way:

> Parents who raise happy, healthy, and successful children create an especially important public good. Children themselves are not the only beneficiaries. Employers profit from access to productive workers. The elderly benefit from Social Security taxes paid by the younger generation … Fellow citizens gain by having productive and law-abiding neighbors. These are all examples of positive spillovers and side effects that economists often call 'positive externalities', because they are external to the actual decision to provide care.[18]

The positive externalities that are produced as a result of child-rearing labor form the basis of a particular kind of pro-sharing argument, known as the *public goods argument*. Put briefly, the idea is that "Parents, by having and rearing children, produce positive externalities for everyone, and non-parents would be unfairly free riding on parents if they accepted the benefits but did not share the costs."[19]

It is important at this point to distinguish between incentive-based and fairness-based versions of the public goods argument.[20] One possible reason for sharing the cost of child-rearing is that the public good that it produces—productive citizens—is in relatively short supply, and people need to be incentivized to produce more of it. Cost sharing is meant to achieve this end by making child-rearing a more attractive, because less costly, option. For example, in countries with declining birth rates, governments have sometimes adopted cost-sharing policies as a means of encouraging population growth, instituting schemes like paid parental leave, universal subsidized daycare, and increased child benefits in the hopes that it will encourage people to have more children. Importantly, these policies are not justified on grounds of fairness, but rather because they are deemed necessary[21] to achieve an optimal population size, e.g., one that is capable of maintaining but not over-burdening existing social welfare systems.

By contrast, another possible reason for sharing the costs of child-rearing is that some form of cost-sharing is required by considerations of fairness. Even when there is no threat of under-supply, non-parents still benefit greatly from the goods created by other people's child-rearing labor and may on that basis have an obligation to share in part of its costs. This fairness-based version of the public goods argument follows from a more general principle of fairness (or "fair play") which states that we should not enjoy the benefits of social cooperation unless we are prepared to do our fair share in creating them. As John Rawls (following H.L.A. Hart) writes,

> ...when a number of persons engage in a mutually advantageous cooperative venture according to rules, and thus restrict their liberty in ways necessary to yield advantages for all, those who have submitted to these restrictions have a right to a similar acquiescence on the part of those who have benefited from their submission. We are not to gain from the cooperative labors of others without doing our fair share.[22]

As a principle governing social cooperation, the principle of fair play has considerable initial appeal, particularly when it is applied to the production of non-excludable[23] public goods like fresh air or a clean environment, as these types of goods are especially susceptible to free rider problems. Imagine, for example, that a group of environmentalists decide to pick up litter on a commonly owned beach, thereby creating a public good.[24] As a result of their service, more people come to use the beach, and it quickly becomes littered with garbage. When it comes time to clean the beach again, it does not seem unreasonable to think that the environmentalists could legitimately demand help from the beachgoers who have enjoyed using it—after all, the beachgoers have benefitted from the good created by the environmentalists' labor, so it seems fair that they should contribute toward the cost of maintaining it.

According to the public goods argument, the same reasoning also applies to the goods that are produced as a result of parents' child-rearing labor. As proponents like Folbre emphasize, non-parents benefit in a number of ways from other people's child-rearing projects: when all goes well, those children grow to become productive workers, law abiding citizens, and perhaps most importantly for the public goods argument, a future tax base on which to maintain the social security programs that will support parents *and* non-parents in old age. Given that non-parents stand to benefit from these goods, considerations of fair play seem to tell in favor of an obligation to share in the cost of producing them, usually through tax contributions that can be redistributed to parents in the form of child subsidies, paid parental leave, and state-sponsored childcare programs.[25]

Is the public goods argument sound? Assuming that child-rearing does in fact create positive externalities (an empirical assumption I will discuss in further detail below), its plausibility ultimately relies on the truth of two premises, namely:

1 the principle of fair play is a defensible principle; and
2 the goods created as a result of parents' child-rearing labor are the kinds of goods that are appropriately governed by the principle of fair play.

Premise (1) has been the subject of considerable debate since the principle of fair play was originally formulated by Hart, with one of the most prominent critiques being advanced by Robert Nozick.[26] According to Nozick, the principle of fair play is implausible because it falsely assumes we can "foist" benefits on people without their consent and then demand that they contribute toward the cost of those benefits. To illustrate his point, Nozick imagines a neighborhood broadcasting system in which each neighbor takes a turn operating a loudspeaker for a day, playing records, reading news bulletins, and telling amusing stories for the enjoyment of anyone who cares to listen. According to Nozick, even if you have occasionally enjoyed listening to the broadcasts of your neighbors, you are not thereby obligated to sacrifice a whole day at the loudspeaker when it comes your turn to operate it—after all, for you, the chore of operating the loudspeaker for a day might dramatically outweigh whatever benefit you received from listening to the broadcasts, casting doubt on the notion that simply receiving a benefit can generate a duty to share in its costs.

However, while Nozick's critique is persuasive in the instance he describes, it is perhaps best interpreted as an argument against a broad interpretation of the fair play principle, rather than against the very idea of fair play obligations.[27] It might seem implausible to claim that passively enjoying a few broadcasts generates a duty to participate in a broadcasting program, though these are not the types of benefits that Hart and Rawls likely had in mind when formulating their principles. In the Rawlsian formulation quoted above, for example, acquiescence on the part of beneficiaries is required only with respect to benefits that have been produced by "mutually advantageous cooperative venture[s]" in which participants have "restrict[ed] their liberty in ways necessary to yield advantages for all." Moreover, it only applies if one "has voluntarily accepted the benefits of the arrangement or taken advantage of the opportunities it offers to further one's interests."[28] Some theorists, such as George Klosko, have narrowed the scope of the principle even further, limiting it to the production of goods that all members of the community can be presumed to want or require.[29] On this view, the principle of fair play only applies to the production of goods that are "necessary for an acceptable life for all members of the community,"[30] not to merely discretionary goods that members can easily do without.

Given these qualifications, a proponent of the fair play principle might respond that while there is no fair play obligation to participate in a public broadcast system, this is because the goods that it produces are not properly governed by the principle of a fair play—either because the broadcast system cannot plausibly be characterized as a mutually advantageous cooperative venture whose benefits have been voluntarily accepted or because the goods that it produces are merely discretionary. However, while the benefits of broadcasting do not seem to fit the bill, there are other cooperatively produced goods that do. National security, for example, is a good that is produced by a mutually advantageous cooperative venture (a political association), and one that is also necessary for an acceptable life for all members of the community.[31] It seems fair, then, that persons who benefit from that good should also contribute toward its costs.

However, most critics of the public goods argument do not question the tenability of (some version of) the principle of fair play, but rather question whether the goods that are produced as a result of parents' child-rearing labor are in fact the kinds of goods whose production is governed by that principle—that is, they have denied premise (2). In fact, three distinct objections have been raised against that premise: one that denies children are positive externalities; one that concedes they are positive externalities, but denies they are like the other public goods that are plausibly governed by the principle of fair play; and one that denies they are public goods at all.

5.3.1 The Negative Externalities Objection

One possible way of challenging the public goods argument is to claim that the empirical premise on which it rests—that child-rearing creates positive externalities—is false. There are in fact two variations of this objection. The first, weaker variation calls attention to the fact that child-rearing does not *always* produce positive externalities; rather, it sometimes produces negative externalities in the form of unjust or unproductive citizens who serve primarily as a drain on social resources. On this view, parents might have a fairness-based claim that non-parents share in the cost of their child-rearing labor, but only to the extent that their labor actually produces a good that non-parents will benefit from.[32] If sound, this weak variation of the negative externalities objection does not serve as an objection to the public goods argument *per se*, but rather as a consideration limiting its scope: it implies that only *some* parents will have fairness-based claims to the cost-sharing contributions of non-parents.

Should we accept this narrower view? On the one hand, it might seem attractive from the perspective of responsibility-sensitive egalitarians, who may see the future productivity of one's child as a matter of good or bad option luck. On this view, having a child is on par with taking a calculated gamble, as parents know beforehand that there is a risk their child will not be socially productive and that they will not be eligible for the cost-sharing contributions of non-parents if that turns out to be the case (though, given that this knowledge is not always available before the child reaches adulthood, this ineligibility might take the form of an obligation to pay back the value of past parental subsidies to the cost-sharing scheme). On the other hand, however, there seem to be good reasons against discriminating between parents on a case-by-case basis. First, and most importantly, there are many cases in which parents cannot meaningfully be held responsible for their children's lack of social productivity. For example, some children will not be socially productive because they are born with, or acquire during their childhood, cognitive or physical impairments that prevent them from social and economic participation, though this outcome is not traceable to any choice on the part of the parents and in fact seems like the kind of bad brute luck that responsibility-sensitive egalitarians are willing to compensate for. Even differences in parenting ability, which may contribute toward variously socially productive children, are not necessarily entirely traceable to the choices of parents, but may be the product of different types of upbringings and other factors that are beyond the parents' control.

Second, there may also be epistemic and pragmatic reasons against discriminating between parents on a case-by-case basis. Perhaps with the exception of children with significant physical or cognitive impairments, it is often difficult to tell which children will be, or have been, net social contributors, and hence which parents are eligible for the cost-sharing contributions of non-parents. These epistemic limitations might be overcome by establishing a system to track and calculate the social productivity of each child, though such a system is not without its own costs. In addition to its significant financial costs, making the requisite calculation of each child's social productivity would inevitably require intrusive interferences in the private lives of parents and children, which may prove to be more trouble than they are worth.

In light of these considerations, many proponents of the public goods argument have preferred to think of parents as collectively producing the good that non-parents will benefit from.[33] On this view, we should not distinguish between parents individually, but should rather treat them as a group of people who have all made the same choice, e.g., the choice to create and raise the next generation of citizens. However, even if we assume that the goods of child-rearing are produced by parents collectively, the public goods argument still faces a stronger version of the negative externalities objection, namely, that in an arguably overpopulated world, the production of *any* children produces negative environmental externalities in virtue of its contribution to overpopulation and anthropogenic climate change.[34] This is a serious objection that may ultimately limit the applicability of the pro-sharing argument, though I will set it aside until Section 5.5. For now, I will assume for the sake of argument that children *do* in fact represent positive externalities in the societies in which they are created and that these are benefits that parents produce collectively.

5.3.2 The Cost/Intentionality Objection

A second possible way of challenging the public goods argument is to show that child-rearing is unlike the other public goods that are governed by the principle of fair play. According to Paula Casal and Andrew Williams, a plausibly construed principle of fair play applies to "a very specific class of positive externalities," namely, to "nonexcludable goods that are produced by *cooperative activity* in which individuals *bear some cost*, which they would not otherwise bear, *in order to produce the good*."[35] In the beach cleaning example illustrated above, we can see that each of these criteria are satisfied: the clean beach is produced as a result of the *cooperative activity* of the environmentalists; it is produced with the *intention* of creating a good for others to enjoy; and its production involves *costs* (in the form of time, effort, and supplies) that the environmentalists would not otherwise have incurred if not for the production of the good. It seems fair, then, that these costs should be shared in some way by the other individuals who have benefitted from using the beach.

However, Casal and Williams argue that the goods produced as a result of parents' child-rearing labor are importantly distinct from the other goods governed by the principle of fair play, as parents do not typically have children with the intention of producing benefits for others, nor do they typically view having children as an all-things-considered cost.[36] In this sense, child-rearing is less like the beach cleaning example described above, and more like a modified beach cleaning example in which a farmer gathers seaweed from a beach for the purpose of fertilizing his fields.[37] In this instance, the farmer's labor happens to create a benefit for others, though given that this is merely a by-product of his self-serving actions, it does not seem like beachgoers would be unfairly free riding on the farmer by enjoying the beach without helping him gather seaweed. In a similar vein, Casal and Williams argue that "merely enjoying the beneficial consequences of others' procreative activity need not violate the principle [of fair play],"[38] as those benefits are not normally produced intentionally, nor does their production involve net costs to the extent that parents themselves benefit from creating and raising children.

Does the cost/intentionality objection succeed in undermining the public goods argument? This depends on (a) whether the incursion of costs and the intentional creation of benefits are in fact necessary for the generation of fair play obligations, and (b) whether parents really do not incur costs and do not intentionally create benefits when pursuing their child-rearing projects. With respect to (a), there may be possible counterexamples that tell against the idea that fair play obligations are *only* generated in cases where individuals incur costs for the sake of intentionally creating benefits. For example, Olsaretti imagines a case in which a group of gardening enthusiasts plant trees as a means of practicing their hobby, which happens to have the beneficial side-effect of offsetting carbon emissions from transport and industry.[39] According to Olsaretti, even if the gardeners did not view tree planting as a burden, and were not motivated by a desire to offset emissions, it still might seem fair that others who stand to benefit from that activity share in its costs. After all, tree planting still involves time, effort, and supplies, and if such an activity was deemed necessary to produce a particular type of good (e.g., clean air), then fairness might demand that the beneficiaries of that good contribute toward its production, *even if* the gardeners would have produced it anyway. Of course, one might still resist this conclusion on cost/intentionality grounds, though it is worth noting that this type of resistance might end up committing one to some rather unattractive implications. For example, if the gardening enthusiasts do not have a legitimate claim to compensation on cost/intentionality grounds, then a teacher who loves her job, and who is motivated primarily by the joy she gets from teaching students, might also have no legitimate claim to compensation for her labor.

However, irrespective of whether this type of counterexample holds, one might also claim in response to (b) that child-rearing *does* in fact exhibit the features of cost and intentionality that Casal and Williams claim are necessary for the generation of fair play obligations, albeit on a less demanding interpretation of what those features require. Regarding the feature of cost, while it is true that parents do not typically view child-rearing as an all-things-considered cost, it is not clear that the *perception* of costs is what is relevant to the generation of fair play obligations; rather, what seems relevant to the generation of those obligations is the actual *incursion* of costs, and parents undeniably satisfy this condition in virtue of the significant amount of resources they invest in the raising of their children.[40] One might object at this point that parents also stand to benefit from these costs in the same way as a normal investment, though this should not affect their claim that costs be shared by other beneficiaries, as it is often the case that the producers of public goods also benefit from their production. For instance, the producers of a system of national security are presumably better off than they otherwise would be in the absence of such a system, though it still seems like they have a legitimate claim that others not free ride on their cooperative efforts.[41]

Regarding the feature of intentionality, Olsaretti suggests that Casal and Williams may be conflating intention with parental motivation when formulating their critique.[42] According to this line of response, while it is true that the creation of benefits for others is not typically what *motivates* parents to have children, it is not true that parents do not *intend* to create the benefit, as much parental labor is explicitly directed toward ensuring that children grow to become productive members of society who are positioned to contribute to redistributive schemes through taxation. One might object that there is a difference between intentionally creating something that is beneficial and intending to benefit others, though it is unclear what effect this difference should have on our obligations of fairness. By Casal and Williams' own estimation, fair play obligations are generated when persons incur costs as a result of intentionally participating in a benefits-producing scheme, so to the extent that parents incur costs to intentionally create productive citizens, then, irrespective of their motivation for doing so, it appears that their labor fits the relevant criteria.

5.3.3 *The Socialized Goods Objection*

However, while the cost/intentionality objection is not entirely persuasive, the public goods argument faces an even stronger objection, which is that the goods produced as a result of parents' child-rearing labor are not in fact public goods at all. Standard examples of public goods like national security or a clean environment tend to exhibit three important features that distinguish them as public goods. First, they are *non-excludable*, in the sense that it is either impossible or too costly to prevent non-contributors from enjoying them. Second, they are usually *unavoidable*, in the sense that they are either impossible or too costly to refuse. Finally, they are also typically *non-rivalrous*, in the sense that one person's enjoyment of the good in question does not diminish or compromise another's.[43] Clean air is an example of a good that exhibits all three of these features: it is impossible to exclude non-contributors (e.g., polluters) from breathing it in, people cannot avoid breathing it in, and one person's enjoyment of clean air does not diminish or compromise another's.

However, as Olsaretti has argued, the goods produced as a result of parents' child-rearing labor do not in fact exhibit *any* of these features, despite initial appearances to the contrary.[44] They are not unavoidable, as non-parents could choose to forgo social security benefits funded by the next generation's taxes; they are not non-rivalrous, as the enjoyment of benefits by one person *does* in fact diminish the amount that is available for others; and perhaps most problematically, they are not non-excludable, as it is conceivable that we could devise a two-tiered social welfare system in which the benefits funded by the next generation's taxes are only accessible to parents. According to Olsaretti, these observations reveal that the benefits of child-rearing are not in fact public goods, but rather deliberately socialized goods that are transferred artificially to non-parents via the social and economic institutions of the welfare state. As she writes,

> ...the central benefits created by parents' having and rearing children are not public goods. Rather, children are *socialized* goods: our social and economic institutions *are intentionally so structured as to ensure that* the product of parents' labors is socially beneficial for everyone, including nonparents. These institutions ensure that children are beneficial for everyone, including non-parents, by forcibly directing some benefits, from the next generation, to today's nonparents and parents alike.[45]

According to Olsaretti, the observation that children are not public goods, but rather deliberately socialized goods, could carry one of two implications for the pro-sharing view. First, it might be thought to undermine that view by weakening the initial case for cost sharing. If the benefits of child-rearing are in principle excludable, then proponents of the anti-sharing view might simply insist that we ought to exclude non-parents from benefits, rather than share the cost of child-rearing. After all, if it is unjust to diminish the resource share of non-parents to finance the child-rearing projects of parents, then it would seem similarly unjust to transfer the benefits of those projects to persons who played no role in supporting them. The socialized goods objection might therefore be thought to lend support to the kind of two-tiered social welfare system noted above, in which non-parents are excluded from the benefits funded by the next generation's tax contributions.

Alternatively, given that a two-tiered system is not in place and that the benefits of child-rearing are typically distributed equally between parents and non-parents alike, the socialized goods objection might actually be thought to lend support to the pro-sharing view by filling an important gap in the public goods argument—namely, the understanding of child-rearing as a "mutually advantageous cooperative venture" capable of giving rise to fair play obligations. Recall that the public goods argument relies on the idea that child-rearing is the kind of cooperative benefits-producing scheme upon which it would be unfair to free ride. While this characterization might have initially seemed plausible, there seem to be at least two problems with it upon reflection. First, child-rearing is not really "cooperative" in the same way as other benefits-producing schemes: decisions to create and raise children are not made collectively in anticipation of the social benefits they will bring, but are usually undertaken privately, producing social benefits only as a happy side-effect, and only then against the backdrop of certain institutional conditions. This in turn reveals a second problem, which is that child-rearing is not *necessarily* benefits-producing (at least not in the material sense being discussed here)[46]; rather, it is benefits-producing only because, and to the extent that, social welfare institutions *make* it benefits-producing by forcibly distributing benefits funded by the next generation to parents and non-parents alike. Taken together, these two problems suggest that the standard version of the public goods argument begs the question, as it must assume the very institutional scheme that it argues for.

However, if we simply posit the existence of social welfare institutions as a premise in a pro-sharing argument, then both of these problems disappear, as the institutions themselves *become* the cooperative benefits-producing scheme. When parents create and raise children in a society with a unified social welfare system, they are contributing to a scheme that deliberately creates benefits for others in the form of pensions, social security, and the like.[47] Given that non-parents in such a society stand to benefit from that scheme, the principle of fair play seems to require that they contribute toward its costs, including the cost of raising the next generation of citizens who will maintain it in the future.

Olsaretti believes that this type of pro-sharing argument—the *socialized goods argument*—provides a more robust case for sharing the costs of child-rearing, particularly in the context of liberal egalitarian societies where unified social welfare systems are already in place. However, while the socialized goods argument does seem to provide a stronger alternative to the standard public goods argument, it still faces a significant hurdle that has yet to be addressed. As Olsaretti herself acknowledges, the socialized goods argument is *conditional*: it does not support cost-sharing obligations in cases where the benefits of child-rearing are not already socialized, nor does it establish that they ought to be socialized.[48] Olsaretti does not see this as an especially pressing problem, as her goal is specifically to determine whether liberal egalitarian societies ought to offer support to families raising children (and she assumes that liberal egalitarians are already committed to the kind of social welfare institutions that can give rise to cost-sharing obligations). However, if we do not take these institutions as a starting point, the conditionality of the argument may be problematic, as critics of the pro-sharing view could still reject the legitimacy of the institutions themselves. Thus, in order to successfully defend a forward-looking pro-sharing argument, it is necessary to explain why we ought to endorse the kind of unified social welfare system on which that type of argument is premised. I undertake this task in the following section.

5.4 Why Socialize the Benefits of Child-Rearing?

Before doing so, however, it is worth pausing for a moment to take stock of the argument so far. One important insight that can be gleaned from the previous two sections is that the answer we provide to the question of parental justice cannot be divorced from the answers we provide to more general questions of distributive justice. What fairness requires vis-à-vis sharing the cost of child-rearing will ultimately depend on what justice requires vis-à-vis a society's distribution of resources. For example, we saw in the previous section that the principle of fair play can support cost-sharing obligations, but only when there is a unified social welfare system in place that translates child-rearing labor into benefits for non-parents. In this sense, the success of the pro-sharing view is conditional on the justice of a particular background distribution: *if* justice requires a unified social welfare system in which the benefits funded by the next generation are distributed to parents and non-parents alike, *then* non-parents have a fairness-based obligation to share in the cost of child-rearing. The same point applies to the anti-sharing view as well. Just as the pro-sharing view must assume an institutional scheme that translates child-rearing labor into benefits for non-parents, the anti-sharing view must assume the absence of such a scheme, for in order for cost-sharing to represent a net burden to non-parents, it must be the case that they do not materially benefit from child-rearing over the course of their lives—a condition that is unlikely to obtain if they are the recipients of social security benefits in old age. The lesson to be taken away, then, is that different distributive schemes generate different kinds of cost-sharing obligations, so the conclusion we reach about how the costs of child-rearing should be distributed throughout society will ultimately depend on what we think about how resources should be distributed more generally.

For some commentators, this is all to the credit of the pro-sharing view, for the fact that the anti-sharing view conflicts with existing social welfare institutions might be read as a *reductio ad absurdum* for that view. This is the perspective taken by Olsaretti, for example, who maintains that "defenders of the anti-sharing view cannot occupy a comfortable middle ground, one in which there is familiar welfare state in place that serves egalitarian justice, but in which nonparents are not obliged to share the costs of children."[49] Olsaretti is correct about the dilemma facing the anti-sharing view, though simply pointing it out does not necessarily constitute an argument against that view, as proponents could still reject the social welfare institutions that give rise to cost-sharing obligations (as some have been inclined to do).[50] Thus, in order to adequately defend the pro-sharing view, we need to provide an independent argument for why we should endorse the kind of unified social welfare system on which that view is premised.

Following Rawls, one way to provide support for such a system is to show that it would be chosen by rational and suitably motivated individuals in a position of initial equality.[51] To get a sense of the shape this argument might take, let us return to the *Desert Island* scenario with which we began. We now know that the real choice faced by the islanders is not the choice between sharing or not sharing the costs of child-rearing within the context of fixed institutions, but rather the choice between different institutional schemes that will generate different kinds of cost-sharing obligations. While there are numerous arrangements from which the islanders might choose, ranging from those that completely internalize the benefits of child-rearing to those that completely externalize them, it will be useful for expository purposes to imagine that their choice is between the two extremes at either end of the spectrum. Thus, upon arriving on the island, let us assume that the islanders face a choice between two institutional schemes for distributing the benefits of child-rearing:

1 a unified social welfare system, in which the costs *and* the benefits of child-rearing are socialized among parents and non-parents alike; or
2 a two-tiered social welfare system, in which both the costs *and* the benefits of child-rearing are internalized by parents alone.

According to Rawlsian contractualism, the best way to determine which scheme is preferable from the perspective of justice is to imagine which would be chosen in a hypothetical choice scenario characterized by fairness and equality. To do this, we imagine that the islanders are making their choice from behind a veil of ignorance that shields them from potentially biasing information. Behind the veil of ignorance, islanders have access to general information about how resources and opportunities would be distributed under either scheme, but they do not have access to personal information that would shed light on their place within that distribution. In particular, they do not know whether they are adults or children, whether their conception of the good includes parenting or remaining childless, and whether they are relatively fortunate or unfortunate in the distribution of natural talents and abilities.

The immediate effect of the veil of ignorance is to prevent the islanders from tailoring the choice of institutions to their own personal circumstances. Equipped with full information about their talents and preferences, islanders might be tempted to select the institutional scheme that works best for them individually, regardless of the effects that it might have on others. However, the veil of ignorance prevents them from doing so. Rather than viewing the choice of institutions from their own individual perspective, each islander is forced to view it from the perspective of every social position that is possible under either scheme, as each of them could end up occupying *any* of these positions. The resulting decision is then thought to be fair, as it is the product of an impartial process in which each possible social position is given equal and non-biased consideration.

Supposing the islanders are mutually disinterested, concerned primarily with maximizing their own resources and opportunities over the course of their lifetime, Rawls believes that the veil of ignorance would lead them to employ a "maximin" decision-making principle in their deliberations, or to select the scheme that optimizes the worst outcome.[52] There are two main reasons for this. The first is because the veil of ignorance shields them from probability estimates that might otherwise inform calculated gambles. For example, suppose that scheme A had a worse worst position than scheme B, but a much better best position. If the parties to the decision knew that there was only a slight chance of ending up in the worst position under scheme A, they might accept that risk in exchange for the much higher payoff of ending up in scheme A's best position. However, because they do not have this information, parties are forced to assign an equal probability to each outcome, which introduces a level of risk aversion into their decision-making. For all they know, there is an equal chance of ending up in the best and worst position under either scheme, so the rational choice is to select the scheme that maximizes the minimum possible return.

The second reason why maximin is an appropriate decision-making principle in this context has to do with the high stakes involved with the decision in question. Rawls concedes that maximin is not a rational decision-making principle for every decision, particularly for those in which potential losses can be recouped in the future, though the choice between different social institutions is not this kind of decision. In choosing between a unified or two-tiered social welfare system, the islanders are making a choice between different background conditions for all of their future choices, one that will have far-reaching and irreversible effects on their future life prospects. Given the unique importance of this decision, Rawls believes that it is rational for parties to approach it conservatively and try to optimize the worst possible outcome.

If Rawls is correct that the islanders would be motivated to select the scheme with the best worst outcome, as measured by one's level of resources and opportunities over the course of a lifetime, which scheme would they choose? Given that the decision is between institutions that will determine the distribution of child-rearing costs, it seems like they would want to evaluate it from at least three different perspectives, corresponding to the three main positions they might find themselves in once the veil has been lifted: an adult with a child-rearing preference ("parent"), an adult with no child-rearing preference ("non-parent"), and a child. Each position can be subdivided further according to whether the individual occupying it is relatively fortunate or unfortunate in terms of the various factors that will influence their resources and opportunities over time. For adults, this includes their level of marketable talent during their working years, their ability to save for retirement, and their option luck in matters regarding childbearing (e.g., whether their child is relatively inexpensive or expensive

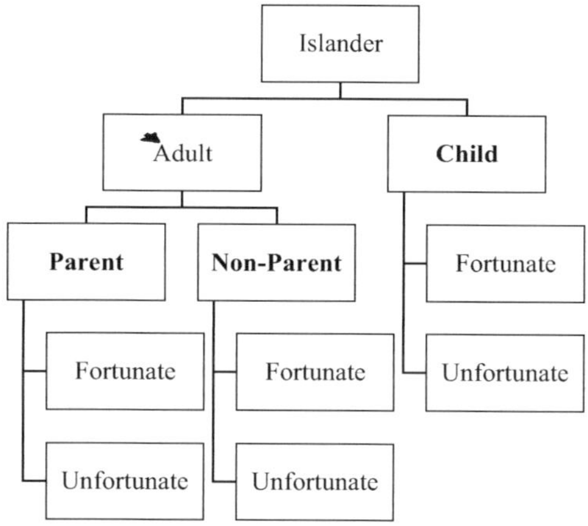

Figure 5.1 Possible social positions on *Desert Island.*

to raise, whether their contraception succeeds or fails, etc.), whereas for children, this mainly includes who their parents are and how expensive they are to raise. The various possibilities are illustrated in Figure 5.1.

If the islanders were guided by maximin reasoning, they would focus on comparing the effects of each institutional scheme on the resources and opportunities of individuals occupying the bottom position on each branch, i.e., unfortunate parents, unfortunate non-parents, and unfortunate children. It seems clear that individuals in *all* of these positions would have a significant reason to prefer scheme (1) over scheme (2), or a society with a unified social welfare system in which the costs *and* benefits of child-rearing were shared among parents and non-parents alike. To explain:

Unfortunate Parents: Unfortunate parents would reach this conclusion by comparing their level of resources and opportunities during their working years under schemes (1) and (2). Under the scheme (2), unfortunate parents are vulnerable to two outcomes they would want to avoid. The first is the inability to fulfill their child-rearing preference because they cannot afford to provide a child with what they are justly due. This may be because they have insufficient income to support themselves along with another person, or because their child has (or is likely to have) special needs that make them too expensive to raise independently. The second is the ability to fulfill their child-rearing preference, but only at the expense of other opportunities. This is because unfortunate parents will be forced to spend a larger portion of their limited resources on meeting the totality of their child's needs, leaving less available for the pursuit of other projects and ambitions.

Whether or not it is ideal, scheme (1) is at least comparatively better for unfortunate parents on both of these counts. If the costs of child-rearing are shared among parents and non-parents alike, it is *more* likely that unfortunate parents will be able to satisfy their child-rearing preferences and those who are able to will be *more* financially secure and will have *more* resources to dedicate toward their other projects and ambitions, regardless of how they fare in absolute terms. Notice, moreover, that the fact that unfortunate parents will fare better under the scheme (1) during their working years is sufficient to establish their preference for it. Because they will have access to social security benefits under both schemes, consideration of their resources and opportunities in retirement will not play a significant role in their decision.

Unfortunate Non-Parents: The preference of unfortunate non-parents for scheme (1) is less obvious and follows from a slightly more complicated line of reasoning. Initially, it seems like unfortunate non-parents face a simple trade-off in their choice of institutions. Under the scheme (1), they will be worse off during their working years by having to share in the cost of child-rearing, but more secure in their retirement, as they will then be entitled to social security benefits funded by the next generation. Under the scheme (2), they will be better off during their working years by avoiding cost-sharing contributions, but less secure in their retirement, as they will not be entitled to publicly funded social security benefits and will therefore have to rely on whatever they have managed to save during their working years. One might infer from this trade-off that there is no clear institutional preference for unfortunate non-parents, though a preference for scheme (1) begins to emerge once we take into account other kinds of considerations. For instance, even though they do not have a preference for child-rearing, it is important to remember that non-parents of all stripes may still be at risk of incurring child-rearing obligations involuntarily as a result of bad option luck, e.g., if their decision to have sex results in an accidental conception, and abortion and adoption are not available options. If they are guided by maximin reasoning, it will be rational for unfortunate non-parents to select the scheme that minimizes the effect of bad option luck on their resources and opportunities. Because child-rearing is less burdensome under scheme (1) than under scheme (2), scheme (1) will be the preferable option.

Moreover, there are also considerations having to do with maximizing opportunity that might lead unfortunate non-parents to prefer scheme (1) over scheme (2). Paul Bou-Habib has pointed out that preferences for and against having children are typically complex and subject to variation, relying as they do on "different inclinations within us that can change in light of experiences that lie ahead."[53] If parties behind the veil of ignorance know that their initial parenting preference could change over time, they might be inclined to protect themselves against the eventuality of changing their mind about parenting, but being unable to fulfill their newfound parenting preference due to an inability to provide a child with what they are justly due. Scheme (1) provides them with better protection against this eventuality than scheme (2).

Unfortunate Children: The preference of unfortunate children for scheme (1) follows fairly straightforwardly from the preferences of unfortunate parents and unfortunate non-parents. The worst outcome for children under either scheme is being born with special needs that make them expensive to raise to unfortunate adults who have a limited capacity to meet the necessary expenses, whether this is a pair of adults who have voluntarily brought them into existence or a pair of adults who have involuntarily incurred parental obligations due to bad option luck stemming from the decision to have sex. Because both unfortunate parents and unfortunate non-parents fare better financially under scheme (1) than scheme (2), unfortunate children will have a clear preference for scheme (1).

This, of course, is just a sketch of one type of argument that might be made to show why we have reason to prefer a society in which the costs and benefits of child-rearing are socialized among parents and non-parents alike—there may be different arguments available that lead to the same conclusion, and the argument here could be developed in greater detail.[54] If successful, however, this type of argument serves to buttress the pro-sharing view by showing that there are independent reasons to prefer the kind of institutional scheme that gives rise to fairness-based cost-sharing obligations. Some theorists may believe that this is an unnecessary step in an argument for cost-sharing, given that such a scheme is already in place in most Western democracies. However, whether or not it is in fact necessary will depend on the general aims of the argument, and whether it is being pitched at the level of ideal or non-ideal theory. If one's aim is simply to show that, amid our existing social welfare institutions, non-parents have a fairness-based obligation to share in the cost of child-rearing, then it may not be necessary to provide additional argumentation for the legitimacy or preferability of those institutions, as the principle of fair play may still apply regardless of whether we take those institutions to be desirable. However, if one's aim is to show that justice requires that non-parents share in the cost of child-rearing, then it is necessary to explain why we ought to endorse the kind of background conditions that give rise to cost-sharing obligations in the first place. In either case, however, the contractualist argument presented here advances the pro-sharing view, either as a necessary step in a pro-sharing argument pitched at the level of ideal theory or as a supplementary consideration in a pro-sharing argument that takes existing institutions as its starting point.

Of course, once it is established *that* the costs of child-rearing ought to be shared among parents and non-parents alike, two further questions need to be addressed: (1) what is a fair distribution of child-rearing costs between parents and non-parents; and (2) what form should cost-sharing take? I will not attempt to provide definitive answers to these questions, though it is worth addressing each of them in brief. In response to (1), it is clear that non-parents are only required to share in the cost of providing children with what they are justly due—they are not on the hook for the various supererogatory goods the parents typically provide for the children, including birthday parties, piano lessons, or trips to summer camp. Moreover, even with respect to the goods that children are justly due, such as education or medical care, there will be logical limits on how much non-parents are obligated to contribute—they cannot, for example, be expected to contribute more than they expect to get in return, as their obligation to contribute is premised on the assumption that they will benefit all-things-considered. As a general rule of thumb, then, it seems like a plausible distribution will be characterized by a kind of proportionality, whereby the amount the non-parents are expected to contribute will be proportional to the amount they can expect to get in return. What this amount is may vary according to social context, though it is reasonable to think it will be markedly less than what parents are obligated to contribute.

In response to (2), I noted earlier that cost-sharing will most plausibly take the form of tax contributions that can be redistributed to parents in the form of child subsidies, paid parental leave, and state-sponsored childcare programs, though it is worth acknowledging here that this is not the only possible form they can take. In addition to sharing costs via redistributive taxation, non-parents could also share in the cost of child-rearing by sharing directly in child-rearing labor, as in the case of collective child-rearing arrangements like the Israeli *kibbutzim*. This might be attractive for advocates of the non-parental care of children,[55] though considerations of personal autonomy will likely weigh against imposing that kind of obligation on persons who have deliberately chosen not to create and raise children.

5.5 Internal and External Limitations on the Pro-Sharing View

Before concluding, I would like to qualify my argument for sharing the cost of child-rearing by briefly taking note of two important limitations of a forward-looking pro-sharing view. The first type, which I will refer to as *internal* limitations, derives from within the society to which a pro-sharing argument is meant to apply, whereas the second type, which I will refer to as *external* limitations, derives from the global community of societies in which that society is situated. Both impose limits on the conditions under which the pro-sharing argument applies and, particularly in the case of external limitations, whether that argument even applies at all.

5.5.1 Internal Limitations

Suppose that the preceding arguments are sound and that non-parents have a fairness-based obligation to share in the cost of child-rearing when there is a unified social welfare system in place that distributes the benefits funded by the next generation to parents and non-parents alike. It does not necessarily follow from this argument that non-parents have an obligation to share the costs of *any and all* children that parents might produce. Rather, if non-parents are obligated to share in the cost of child-rearing in virtue of their status as the future beneficiaries of parents' child-rearing labor, then they are only obligated to share the cost of producing whatever number of children is necessary to secure those benefits in the future. Calculating this number in real-world contexts is bound to be extraordinarily complicated, as one would have to consider factors such as a society's level of productivity, its cost of living, and its rate of inflation among others, though its general significance can be illustrated with a simplified example. Imagine that a society contains an initial cohort of 100 adults (G1), and the economic conditions are such that G1 will be required to produce the same number of children (G2) in order to maintain the social security system on which it will depend in old age at a reasonable level of taxation. The non-parents of G1 will have a fairness-based obligation to share the cost of producing 100 children, though it is not clear that they have the fairness-based obligation to share the cost of any children exceeding the required 100, nor is it clear that they have a fairness-based obligation to share the cost of producing *any* children should the number of children comprising G2 fall significantly below 100. Let me explain both of these points in further detail.

The reason why the non-parents of G1 do not have a fairness-based obligation to share the cost of any children exceeding the required 100 is that the creation of additional children is not necessary to secure the benefits that give rise to their cost-sharing obligations. Suppose that G1 consists of 50 non-parent individuals and 25 parent couples, each of which produces five children. Assuming that social security benefits are fixed, rather than proportional to the size of G2, the non-parents of G1 will have an obligation to share the cost of the first four children of each couple, though they will not have an obligation to share the cost of the fifth. The point becomes clearer the larger that G2 expands beyond the size that is required to sustain G1 in old age. While five versus four children per couple may not seem like a significant difference, it is harder to argue that non-parents have an obligation to share the cost of eight children per couple if only four are required to produce the future benefits that give rise to their cost-sharing obligations in the first place.

One might claim that we can address this worry by making the social security benefits distributed to members of G1 proportional to the size of

G2, such that non-parents would be entitled to a larger package of social security benefits in virtue of sharing the cost of a larger number of children during their working years. This type of strategy might succeed within a limited range (e.g., 4-*x* children per couple) though there will certainly be limits on how far it can be taken. First, there will be a clear upper limit on the number of children that the members of G1 can sustain at a reasonable level of taxation. Because non-parents will be required to contribute a larger share of resources with every additional child, the proportionality strategy becomes less plausible as a greater value is assigned to *x*. Moreover, if continued indefinitely, the proportionality strategy also faces the risk of a demographic disaster in the distant future. If G2 is significantly larger than G1, then G3 will likely have to be at least as large as G2 to support it in old age; and if G3 is significantly larger than G2, then G4 will likely have to be at least as large as G3 to support it in old age; and so on. In this sense, even if we make social security benefits for members of G1 proportional to the size of G2, there will still limits on the number of children that a cost-sharing scheme can accommodate deriving from both the current cost-sharing capability of G1 and a society's general carrying capacity in the distant future.

At the other end of the spectrum, the reason why non-parents of G1 do not have a fairness-based obligation to share the cost of *any* children should the number of G2 fall significantly below 100 is simply because they do not stand to materially benefit from those children in the future (supposing that G2 must contain at least 100 children to maintain the social security system on which G1 will depend). In this sense, the rate of procreation in a given society must fall within a happy medium in order to generate cost-sharing obligations: there must be enough children created to produce positive externalities, but not so many to produce negative externalities.

There are also a few additional considerations that must be taken into account when thinking about internal limitations, which I will mention only briefly here for the sake of completeness. First, in addition to determining the population range of G2 that is capable of giving rise to cost-sharing obligations, one must also determine how "procreative credits" (for lack of a better term) ought to be distributed among the adults in G1 who have a parenting preference. For example, suppose that in our society of 100 adults, there are 20 non-parent individuals and 40 couples with a parenting preference. One now has to decide how the 100 procreative credits ought to be distributed between the 40 couples. One strategy would be to claim that the parents of the first 100 children born are entitled to have the costs of those children subsidized by contributions from non-parents, though if we take the opportunity to parent to itself be a distribuendum of justice, then a fairer strategy might be to assign two procreative credits to each couple and then distribute the remaining 20 credits fairly among any remaining couples who desire more than two children (perhaps by lottery). Couples would be permitted to create a number of children that exceeds their number of credits, but they would not be entitled to cost-sharing contributions for their extra children.

Moreover, one also has to take into account considerations of intergenerational justice when determining how a cost-sharing scheme will work over time. For example, in our society with 100 adults, intergenerational inequities begin to emerge the further that the size of the following generation departs from replacement-level procreation. If G2 is significantly smaller than G1, then this might place an unfair burden on members of G2, who will be tasked with supporting a much larger G1 in old age. Alternatively, if G2 is significantly larger than G1, then the burden is shifted to G1 (which must share the cost of raising a much larger G2) and potentially G3 (which, depending on its size, might also have to support a much larger G2 in old age). Depending on what distributive principle one endorses with respect to intergenerational justice (e.g., egalitarianism, sufficientarianism, etc.), there may be limits on the rate of procreation that derive from the just claims of members of future generations.

5.5.2 External Limitations

In addition to these internal limitations, a pro-sharing view is also subject to external limitations that derive from the global community in which a particular society is situated. As I mentioned a moment ago, the rate of procreation in a particular society must fall within a happy medium in order to generate cost-sharing obligations, with enough children created to generate positive externalities, but not so many children created to generate negative externalities. When we take a society as a self-contained entity, this means that the number of children in G2 cannot be so large as to place an unfair burden on members of G1 or to lead to an ever-ballooning population in which each generation needs to produce greater and greater numbers of children to sustain itself in old age. However, if we take a society as a member of a global community of societies, then the rate of procreation may also be bounded further by the demands of global distributive justice. For example, in a Western industrialized society with a high level of average lifetime consumption, creating even a sufficient number of children to generate cost-sharing obligations may be violating global duties of justice by contributing disproportionality to climate change and environmental degradation. In other words, given a particular level of consumption, the child-rearing labor of parents in affluent societies may create global negative externalities at the same time as it creates domestic public goods.[56]

Depending on how we conceive of the demands of global justice, this may have one of two implications for the success of a forward-looking pro-sharing argument. First, it might imply that that type of argument simply does not work in societies with a level of average lifetime consumption above a certain threshold, as creating a sufficient number of children to generate cost-sharing obligations would violate global duties of justice, and fulfilling global duties of justice would not permit the creation of a sufficient number of children to generate cost-sharing obligations. Alternatively, it might imply that, for the affluent societies at which it is typically directed, the success of the pro-sharing argument will be contingent on reducing the level of average lifetime consumption to a level that is consistent with the demands of global justice when the rate of procreation is at a level sufficient to generate cost-sharing obligations. This is the more attractive implication, though it may require a radical change in the lifestyle to which citizens of industrialized countries are currently accustomed, or radical technological innovation that would enable them to maintain their current lifestyle with a smaller environmental footprint.

I will address some of these issues further in Chapter 7, though their significance here should not be understated. Participants in the debate on sharing the cost of child-rearing—especially proponents of the pro-sharing view—have sometimes ignored how the demands of global justice might impose limits on the tenability of their arguments. Olsaretti, the most prominent defender of the pro-sharing view, notably sets these considerations aside, claiming that substantiating the crucial premise that children are net positive externalities "would require venturing into complex factual territory."[57] This is certainly true, but the facts cannot be ignored or taken for granted, for when we think about procreation in a global context, it might well be the case that one person's benefit is another person's burden. This should be especially concerning to proponents of the pro-sharing argument who rely on a Rawlsian principle of fairness to make their case, for as Rawls himself emphasizes, the principle of fairness applies principally to *just* schemes of social cooperation.[58] If our duties of justice include the duties of global distributive justice, then it is not obviously clear that a cost-sharing scheme meets this criterion if it creates domestic public goods but global negative externalities.

Notes

1 M. Lino et al. *Expenditures on Children by Families, 2015* (Washington, DC: U.S. Department of Agriculture, 2017), ii.
2 Lino et al. *Expenditures on Children by Families, 2015*, pp. 10–14.
3 Serena Olsaretti. "Children as Public Goods?" *Philosophy and Public Affairs*, 41(3) (2013), pp. 226–258, 229.
4 See Paula Casal. "Environmentalism, Procreation, and the Principle of Fairness," *Public Affairs Quarterly*, 13(4) (1999), pp. 363–376; Paula Casal and Andrew Williams. "Equality of Resources and Procreative Justice" in Justine Burley (ed.) *Dworkin and His Critics* (Malden, MA: Blackwell, 2004); Matthew Clayton.

Justice and Legitimacy in Upbringing (Oxford: Oxford University Press, 2006), esp. 67–70; Eric Rakowski. *Equal Justice* (Oxford: Oxford University Press, 1991), esp. 150–154; and Peter Vallentyne. "Equality and the Duties of Procreators" in David Archard and Colin. M. Macleod (eds.) *The Moral and Political Status of Children* (Oxford: Oxford University Press, 2002).

5 In this chapter, I assume for the sake of argument that childbearing and childrearing are decisions undertaken voluntarily.

6 Nancy Folbre. *The Invisible Heart* (New York: New Press, 2001); Rolf George. "Who Should Bear the Costs of Children?" *Public Affairs Quarterly*, 1(1) (1987), pp. 1–42; and Olsaretti, "Children as Public Goods?".

7 Patrick Tomlin. "Should Kids Pay Their Own Way?" *Political Studies*, 63(3) (2015), pp. 663–678. For a critique of this view, see Erik Magnusson. "Parental Justice and the Kids Pay View," *Ethical Theory and Moral Practice*, 21(4) (2018), pp. 963–977.

8 This type of example—a favorite in the literature surrounding parental justice—is based on one originally developed by Ronald Dworkin to illustrate his theory of distributive justice, "equality of resources." For the original example, see Ronald Dworkin. *Sovereign Virtue: The Theory and Practice of Equality* (Cambridge, MA: Harvard University Press, 2000), 65–71. For the most comprehensive discussion regarding its application to parental justice, see Casal and Williams, "Equality of Resources and Procreative Justice," esp. 152–156.

9 Rakowski, *Equal Justice*, 153. See also Clayton, *Justice and Legitimacy in Upbringing*, 69, and Vallentyne, "Equality and the Duties of Procreators," 207.

10 For a full elaboration of this brand of responsibility-sensitive egalitarianism, see Dworkin, *Sovereign Virtue*, esp. chapters 1 and 2.

11 For a general discussion of the principle of stakes, see Serena Olsaretti. "Responsibility and the Consequences of Choice," *Proceedings of the Aristotelian Society*, 109 (2009), pp. 165–188. For a discussion of its significance for the question of parental justice, see "Choice, Circumstance, and the Cost of Children" in Steven de Wijze, Matthew H. Kramer, and Ian Carter (eds.) *Hillel Steiner and the Anatomy of Justice: Themes and Challenges* (New York: Routledge, 2009).

12 For example, while Rakowski argues that procreators are responsible for paying the costs associated with providing their children with what they are justly due (interpreted as an equal bundle of resources), Peter Vallentyne has put forth a far more demanding view, according to which procreators are also liable for paying the costs of children's bad brute luck and any injustices they may inflict on others throughout their lifetime. See Vallentyne, "Equality and the Duties of Procreators."

13 For further discussion of the distinction between "costs of care" and "costs of added adult members," see Olsaretti, "Children as Public Goods?" 229–232.

14 Tomlin, "Should Kids Pay Their Own Way?" 665.

15 G.A. Cohen. "On the Currency of Egalitarian Justice," *Ethics*, 99(4) (1989), pp. 906–944, esp. 921–924.

16 Justine Burley considers but ultimately rejects this type of argument in "The Price of Eggs: Who Should Bear the Costs of Fertility Treatment?" in John Harris and Soren Holm (eds.) *The Future of Human Reproduction* (Oxford: Oxford University Press, 1998), 137–141.

17 For a discussion of the implications of liberal neutrality for the issue of parental subsidies, see Robert S. Taylor. "Children as Projects and Persons: A Liberal Antinomy," *Social Theory and Practice*, 35(4) (2009), pp. 555–576. For a

similar argument in the context of state-funded fertility treatment, see Emily McTernan. "Should Fertility Treatment Be State Funded?" *Journal of Applied Philosophy*, 32(3) (2015), pp. 227–240.
18 Folbre, *The Invisible Heart*, 50.
19 Olsaretti, "Children as Public Goods? 228–229.
20 Casal, "Environmentalism, Procreation, and the Principle of Fairness," 366; Olsaretti, "Children as Public Goods?" 237–238.
21 Perhaps falsely so: states can also achieve an optimal population size by augmenting low fertility rates with an increase in immigration.
22 John Rawls. *A Theory of Justice* (Cambridge, MA: Harvard University Press, 1971), 112. For H.L.A. Hart's original formulation, see H.L.A Hart. "Are There Any Natural Rights?" in Jeremy Waldron (ed.) *Theories of Rights* (Oxford: Oxford University Press, 1984), 85.
23 A non-excludable good is a good that is impossible or too costly to prevent others from enjoying. I return to this feature in further detail below.
24 Casal, "Environmentalism, Procreation, and the Principle of Fairness," 365.
25 It is worth noting here that redistributive taxation is not the only way that child-rearing costs can be shared among parents and non-parents. For example, one alternative way to share the cost of child-rearing might be to require non-parents to perform a certain proportion of child-rearing labor (as in Israeli *kibbutzim*), which may be attractive for advocates of the non-parental care of children. While I acknowledge that there are alternative cost-sharing arrangements to redistributive taxation, I do not consider these in any detail here on the grounds that we can separate the question of *whether* non-parents are obligated to share in the costs of child-rearing from the question of *how* those costs should be shared.
26 Robert Nozick. *Anarchy, State, and Utopia* (New York: Basic Books, 1974), 90–95.
27 For an argument to this effect, see Richard J. Arneson. "The Principle of Fairness and Free-Rider Problems," *Ethics*, 92(4) (1982), pp. 616–633.
28 Rawls, *A Theory of Justice*, 112.
29 George Klosko. "Presumptive Benefit, Fairness, and Political Obligation," *Philosophy and Public Affairs*, 16(3) (1987), pp. 241–259.
30 Klosko, "Presumptive Benefit, Fairness, and Political Obligation," 246.
31 This example, while commonly invoked in favor of the fair play principle, is not uncontroversial. Some theorists, most notably A. John Simmons, have argued that fair play obligations apply only to goods that one enjoys but is able to refuse, indicating a distinction between the active acceptance and passive receipt of benefits. See A. John Simmons. "The Principle of Fair Play," *Philosophy and Public Affairs*, 8(4) (1979), pp. 307–337.
32 Vallentyne, "Equality and the Duties of Procreators."
33 Olsaretti, "Children as Public Goods?" 235–237.
34 Casal, "Environmentalism, Procreation, and the Principle of Fairness," 370–373.
35 Casal and Williams, "Equality of Resources and Procreative Justice," 158–159. Emphasis in original.
36 Casal and Williams, "Equality of Resources and Procreative Justice," 159. There is a further concern that child-rearing cannot plausibly be construed as a cooperative activity. I address this concern in the following subsection.
37 Casal, "Environmentalism, Procreation, and the Principle of Fairness," 368.
38 Casal and Williams, "Equality of Resources and Procreative Justice," 159.
39 Olsaretti, "Children as Public Goods?", 243.

40 For similar arguments, see Olsaretti, "Children as Public Goods?", 242–245; and Tomlin, "Should Kids Pay their Own Way?" 670.
41 In fact, the idea of producers benefitting seems to be assumed in the very idea of something's being a public good, for if the cost of producing the good outweighed its benefits—such that its production was viewed as a net cost—then it would not be a good at all.
42 Olsaretti, "Children as Public Goods?" 245–248.
43 See, for example, Arneson, "The Principle of Fairness and Free Rider Problems," 618.
44 Olsaretti, "Children as Public Goods?" 248–253.
45 Olsaretti, "Children as Public Goods?" 252. Emphasis in original.
46 While child-rearing is not necessarily benefits-producing in a material sense outside of redistributive social welfare institutions, there may be non-material benefits associated with child-rearing that are produced independent of those institutions. Samuel Scheffler, for example, has argued that the value of many life projects is predicated on the assumption of a "collective afterlife," or the idea that future generations will live on after us. If sound, this argument might show how creating and raising children produces benefits for non-parents even if they are not benefitted materially through social security. See Samuel Scheffler. *Death and the Afterlife*. Ed. Niko Kolodny (Oxford: Oxford University Press, 2013).
47 There is an important caveat to this claim having to do with the rate of procreation, which I will discuss in further detail in Section 5.5.
48 Oslaretti, "Children as Public Goods?" 254.
49 Olsaretti, "Children as Public Goods?" 255.
50 Hillel Steiner and Peter Vallentyne might be read as endorsing this conclusion. See Hillel Steiner and Peter Vallentyne. "Libertarian Theories of Intergenerational Justice" in Axel Gosseries and Lukas H. Meyer (eds.) *Intergenerational Justice* (Oxford: Oxford University Press, 2009), esp. 66–68.
51 Rawls, *A Theory of Justice*, 1971.
52 Rawls, *A Theory of Justice*, 152–157.
53 Paul Bou-Habib. "Parental Subsidies: The Argument from Insurance," *Politics, Philosophy, and Economics*, 12(2), pp. 197–216, 209.
54 One alternative proposed by Bou-Habib appeals to a Dworkinian hypothetical insurance scheme in which we ask what types of outcomes people would rationally insure against. It should be noted, however, that Bou-Habib does not deploy this scheme as a supplementary argument in a socialized goods account, but rather as a separate account that supports similar pro-sharing conclusions directly. See Bou-Habib, "Parental Subsidies."
55 See, for example, Anca Gheaus. "Arguments for Nonparental Care for Children," *Social Theory and Practice*, 37(3) (2011), pp. 483–509.
56 Interestingly, the existing literature on sharing the costs of child-rearing contains very little discussion about how the demands of global or environmental justice may serve to temper pro-sharing arguments. For a notable exception, see Casal, "Environmentalism, Procreation, and the Principle of Fairness," esp. 370–373.
57 Olsaretti, "Children as Public Goods?" 239.
58 Particularly, Rawls states that "[The principle of fairness] holds that a person is required to do his part as defined by the rules of an institution when two conditions are met: first, the institution is just (or fair), that is it satisfies the two principles of justice; and second, one has voluntarily accepted the benefits of the arrangement or taken advantage of the opportunities it offers to further one's interests." See Rawls, *A Theory of Justice*, 111–112.

6 Procreation and Adoption

6.1 Introduction

In previous chapters, I have argued that procreators have a duty to only create children in conditions in which their basic welfare interests are likely to be satisfied for the duration of their childhoods, including their interest in adequate parental care, and that being morally responsible for a child's existence generates an inalienable obligation to ensure that the child is adequately parented, either by parenting the child oneself or by delegating caretaking responsibilities to a capable and willing volunteer. When both of these duties are fulfilled, children are born into conditions in which they have access to the material and relational goods they require for their healthy development, thus providing the basis for what I described in Chapter 2 as a "decent start in life." Unfortunately, however, due to both choice and circumstance, there are many instances in which these duties go unfulfilled, leading to the existence of large numbers of children who chronically lack access to adequate parental care. While a precise number is notoriously difficult to calculate, it is estimated that there are currently over 16 million orphans worldwide, where an orphan is defined as a child who has lost or been relinquished by both of their custodial parents.[1] Of this 16 million, between 2 and 8 million are in the care of institutions, while the remainder float precariously between alternative childrearing arrangements, including group homes, foster homes, and temporary care arrangements with relatives or members of their communities. Notably excluded from these estimates are potentially millions more children for whom there is no reliable documentation, including those in unregistered institutions, those living on the street, and those who are at risk of becoming parentless in the near future due to famine, disease, or political violence. Once we factor these children into our estimate, the total number of children currently or soon to be in need of parental care is potentially in the tens of millions.[2]

DOI: 10.4324/9781003164104-7

According to Tina Rulli, these numbers are indicative of a *global orphan crisis*.[3] This crisis is characterized not only by its great magnitude, in the sense that there are millions of children currently in need of parental care, but also by its great severity, in the sense that the cost to children of going without parental care is often devastating.[4] Children who are living on the street or in institutions often lack access to basic nutrition and medical care, and they normally face a far higher risk of neglect, abuse, and exploitation than children who are living with families. Even children who are raised in seemingly preferable arrangements, such as group homes or foster homes, often suffer a wide range of developmental problems due to a lack of continuous care and attention, including attachment disorders, impaired intelligence, and long-term problems with executive function and social interaction.[5] Not surprisingly, the future life prospects for many orphaned children are accordingly grim, with many succumbing to a range of deficits by the time they reach adulthood, including poverty, homelessness, and unemployment.

Prospective parents could help to avert this fate by adopting orphans into their families and providing them with long-term parental care, yet most choose to bring new children into existence instead. Is this morally justifiable? Several philosophers have recently considered the claim that it is wrong to create new children when there are so many existing children in need of parental care.[6] According to this view, we have a significant moral reason to meet the needs of existing children before creating additional needs to be met, and this reason is not easily defeated by competing considerations, including individual preferences for biological children or the substantial costs that are often associated with adopting a child. If sound, this type of argument has significant implications for the morality of procreation, suggesting that as long as there are existing children in need of parental care, it is often morally wrong to bring additional children into existence.

In this chapter, I critically examine the most prominent version of this argument in the existing literature, namely, the *rescue-based argument*, which derives an individual duty to adopt from a more general duty to rescue or assist those in need. Despite its initial plausibility and noble intentions, I argue that it rests on two questionable premises, namely, (1) that we are morally obligated to prevent serious harm to others so long as we can do so at comparatively little cost to ourselves and (2) that prospective parents are able to prevent serious harm to orphaned children at comparatively little cost to themselves by adopting them into their families and providing them with long-term parental care. The first premise is not necessarily true in cases that resemble the orphan crisis, where one's intervention is merely sufficient to prevent serious harm to a victim, whereas the second premise is often false given the significant financial, emotional, and agency-related costs of adoption, particularly in current, non-ideal conditions. The upshot of these limitations is that the rescue-based argument can only generate a duty to adopt with respect to a relatively small constituency of orphaned children, many of whom are likely to be adopted anyway by willing volunteers. In the vast majority of cases, then, the rescue-based argument does not entail that it is wrong to create new children when there are already existing children in need of parental care.

The chapter proceeds in four parts. In Section 6.2 I outline the rescue-based argument for the duty to adopt. In Sections 6.3 and 6.4, I advance two lines of objection to this argument: one that interrogates the conditions under which we have a duty to rescue, and one that interrogates whether the most plausible interpretation of the duty to rescue can generate an individual duty to adopt. Finally, in Section 6.5, I consider the possibility of collective obligations to facilitate adoption through the provision of financial support to adoptive parents.

6.2 The Rescue-Based Argument for the Duty to Adopt

In the existing literature surrounding the ethics of adoption, the duty to adopt has typically been derived from a more general duty to rescue or assist those in need, as famously illustrated in cases like *Shallow Pond*:

> *Shallow Pond*: The path from the library at your university to the humanities lecture hall passes a shallow ornamental pond. On your way to give a lecture, you notice that a small child has fallen in and is in danger of drowning. If you wade in and pull the child out, it will mean getting your clothes muddy and either cancelling your lecture or delaying it until you can find something clean and dry to wear. If you pass by the child, then, while you'll give your lecture on time, the child will die straightaway. You pass by and, as expected, the child dies.[7]

It is clearly wrong for you to proceed to a lecture while a child helplessly drowns in a nearby pond, and the reasoning for this is straightforward: there is a child in danger of losing their life, and this loss can be easily prevented at comparatively little cost to yourself. Peter Singer translates this type of reasoning into a generalized moral principle, which I will paraphrase here as the *duty to rescue*:

> *Duty to Rescue*: If we can prevent serious harm to others at comparatively little cost to ourselves, we have a moral duty to do so.[8]

Daniel Friedrich and Tina Rulli have independently argued that accepting a duty to rescue in cases like *Shallow Pond* commits us to a strong *pro tanto* duty to adopt, as the features that generate a duty to rescue in *Shallow Pond* are also present in the case of the orphan crisis.[9] First, like the child in the pond, orphaned children are also at risk of losing what Rulli calls "critical life goods," or goods that are "necessary for a person to avoid being in a state of critical, absolute harm."[10] Not every orphan faces an imminent risk of death, though many face a risk of significant physical and psychological harm as a result of lacking continuous and long-term parental care. Common physical harms include malnutrition, poor health, and greater susceptibility to neglect, abuse, and exploitation, while common psychological harms include impaired cognitive and emotional development due to a lack of continuous care and attention.[11] Given its detrimental effects on a child's present well-being and future life prospects, Rulli argues that a childhood bereft of adequate parental care is plausibly characterized as "a severe life disadvantage and impairment."[12]

Second, the orphan crisis also features numerous potential rescuers—prospective parents—who are able to prevent this harm at comparatively little cost to themselves by adopting orphans into their families and providing them with long-term parental care. This is not to suggest that adopting children is costless, for as proponents of the rescue-based argument fully acknowledge, adoption often involves a range of financial, emotional, and agency-related costs that place considerable burdens on adoptive parents, particularly in current, non-ideal conditions. However, the important point for the analogy is that the orphan crisis exhibits similar *cost comparisons* as standard rescue cases. Whether or not we think it is too costly in absolute terms, the cost to a prospective parent of rescuing an orphan via adoption is far less than the cost to the orphan of *not* being rescued, much like the cost to a passerby of pulling a child out of a pond is far less than the cost to the child of *not* being pulled out. Given these similarities, both Rulli and Friedrich conclude that insofar as we acknowledge a duty to rescue in cases like *Shallow Pond*, we should also acknowledge a duty to adopt in the structurally similar case of the global orphan crisis.

There are at least two important features of the rescue-based argument that are worth taking note of. First, both Rulli and Friedrich interpret the duty to adopt as a *pro tanto* duty that may be overridden in light of competing moral considerations. One important implication of this interpretation is that the duty itself applies only to *prospective parents* or those who already have an interest in parenting a child. If applied to prospective *non-parents*, or those who do not have an interest in parenting a child, the duty to adopt would involve incurring not only whatever costs are associated with becoming a parent via adoption rather than procreation but also whatever costs are associated with becoming a parent rather than remaining childless, including the opportunity costs associated with forgoing other valuable projects and aims. Both Rulli and Friedrich believe that the latter costs are too demanding to be morally required, and hence that the rescue-based argument only generates a duty to adopt among prospective parents. The proposed duty is thus more accurately described as *the duty to adopt rather than procreate as a means of fulfilling one's interest in parenting*, though for the sake of brevity, I will continue to refer to this duty as *the duty to adopt* in the remainder of the chapter.

Second, both Rulli and Friedrich also assume that preventing the relevant form of harm to orphaned children can only be achieved by adopting them into families and providing them with long-term parental care. Measures that fall short of this ideal, including the provision of high-quality foster or institutional care, are regarded as insufficient. This assumption is significant because it limits the extent to which the costs associated with harm prevention can be divided and distributed across multiple prospective duty-bearers. In some rescue cases, it is possible to divide the costs of assistance into fair shares that can be distributed across multiple prospective duty-bearers, such as in cases of poverty relief in which the relevant form of assistance required is financial.[13] For Rulli and Friedrich, however, the orphan crisis is not one of these cases. While parenting a child does involve financial costs that can be divided to a certain extent among society's members, the "parental resource," as Rulli calls it, also involves a bundle of goods that are indivisible, such as "long-term commitment, time, effort, and emotional care."[14] As a non-liquid resource, it cannot necessarily be purchased or easily divided among society's members, implying that the costs associated with its provision must primarily be borne by individuals.

With these clarifications in mind, is the rescue-based argument sound? Consider its structure:

1 If we can prevent serious harm to others at comparatively little cost to ourselves, we have a moral duty to do so;
2 Orphaned children are at risk of serious harm as a result of lacking access to long-term parental care;
3 Prospective parents can prevent this harm at comparatively little cost to themselves by adopting rather than procreating as a means of fulfilling their interest in parenting; therefore,
4 Prospective parents have a moral duty to adopt rather than procreate as a means of fulfilling their interest in parenting.

I assume this argument is formally valid, and that premise (2) is uncontroversially true based on the available empirical evidence; thus, the soundness of this argument will ultimately turn on the truth of premises (1) and (3). In what follows, I present objections to both of these premises, which, taken together, place considerable limits on the cases in which prospective parents are under a moral duty to adopt rather than procreate.

6.3 Under What Conditions Do We Have a Duty to Rescue?

If we acknowledge a duty to rescue the child from the pond, then we acknowledge a duty to make *some* personal sacrifice in order to prevent serious harm to others, and hence some version of premise (1) must be true. However, the general plausibility of this premise, as well as its ability to generate a duty to adopt, will depend crucially on the upper limit of the costs we are required to incur in order to prevent serious harm to others, or what is meant by the phrase "at comparatively little cost to ourselves." It is useful, then, to distinguish between *strong*, *weak*, and *moderate* interpretations:

Strong: we are morally required to prevent serious harm to others as long as we can do so at a cost that is lower than the harm that is to be prevented by our intervention.
Weak: we are morally required to prevent serious harm to others as long as we can do so at minimal cost to ourselves.
Moderate: we are morally required to prevent serious harm to others as long as we can do so without incurring major costs to ourselves.

Each of these interpretations will require us to rescue the child from the pond, though each will have dramatically different implications for other cases in which we can prevent serious harm to others, including the global orphan crisis. Which, then, should proponents of the rescue-based argument endorse? This will depend on both the general plausibility of the principle reflected in each interpretation and its implications in the context of the rescue-based argument.

A *strong* interpretation of premise (1) requires that we prevent serious harm to others as long as we can do so at a cost that is lower than the harm that is to be prevented by our intervention. This is the interpretation favored by Singer, for example, who argues that we are morally obligated to assist the needy "until we reach the point of marginal utility—that is, the level at which, by giving more, I would cause more suffering to myself or my dependents than I would relieve by my gift."[15] Not surprisingly, many have objected to the *demandingness* of this requirement, arguing that it is inconsistent with living an autonomous life characterized by valuable projects or relationships.[16] For example, most people would not consider it morally impermissible to play recreational soccer or to save for your children's education, though both may run afoul of a strong interpretation of the duty to rescue if one's time and resources could have been used to provide critical benefits to those in need. To the extent that we take this implication to be implausible, we may be led to reject a strong interpretation of premise (1).

Proponents of a strong interpretation are unlikely to be moved by this objection—Singer's point, after all, is that morality is more demanding than we think and hence that we often, perhaps usually, fail to live up to its demands. However, in addition to the general problem of demandingness, a strong interpretation of premise (1) also faces a number of more specific problems in the context of the rescue-based argument that should lead proponents to reject it as a tenable interpretation. First, notice that a strong interpretation of premise (1) is inconsistent with restricting the duty to adopt to prospective parents, insofar as the cost to a non-parent of becoming a parent is lower than the cost to a child of going without parental care. In this sense, a strong interpretation of premise (1) may end up supporting a moral obligation to *become* a parent, which proponents of the rescue-based argument generally regard as implausible. Second, once there are no more children left to adopt, a strong interpretation of premise (1) may also lead to a limited and conditional form of anti-natalism, for if we are morally obligated to prevent serious harm to others to the point of marginal utility, then it is difficult to see how one could justify investing the significant amount of time and resources required to create and raise new human beings when there are so many existing human beings who could benefit from one's assistance.[17] Morally permissible procreation would then be limited to whatever level is necessary to prevent serious harm to presently existing people, who depend in various ways on future generations living on after them.[18]

A *weak* interpretation of premise (1) avoids these problems by setting a lower upper limit for the costs we are required to incur to prevent serious harm to others, though in so doing invites its own set of problems. As a general principle, a weak interpretation of premise (1) is vulnerable to the problem of *egoism* by attributing too much weight to an agent's own self-interest and too little weight to the interests of others. Imagine, for example, that the costs associated with rescuing the child in *Shallow Pond* exceed what we might intuitively regard as a minimal cost threshold: in addition to getting your clothes muddy and missing your lecture, you will also miss out on your $10,000 speaking fee. A weak interpretation of premise (1) may not recognize a duty to rescue the child in this case, but this is implausible. The child's interest in being rescued is clearly weightier from a moral point of view than your interest in collecting the speaking fee, and any principle that fails to register this weighting ought to be rejected. Notice, moreover, that a weak interpretation of premise (1) is generally unavailable to proponents of the rescue-based argument, for the process of adopting a child is likely to impose a variety of risks, costs, and burdens that exceed a minimal cost threshold, in which case premises (3) and (4) of the rescue-based argument would not follow. In this sense, a weak interpretation of premise (1) will simply not generate a duty to adopt.

We seem to be left, then, with a *moderate* interpretation of premise (1), according to which we are morally required to prevent serious harm to others so long as we can do so without incurring major costs to ourselves. At first glance, a moderate interpretation seems capable of striking an intuitive balance between the excessive demandingness of the strong interpretation and the excessive egoism of the weak interpretation, affirming a duty to rescue in a significant range of cases without requiring us to prevent serious harm to the point of marginal utility. Moreover, within the context of the rescue-based argument, it also seems demanding enough to generate a duty to adopt given the non-negligible costs of adoption, but not so demanding that it commits proponents to claims they might otherwise reject, such as a moral obligation to become a parent or a limited and conditional form of anti-natalism.

Of course, whether the moderate interpretation can actually deliver these advantages will depend on how we understand the notion of a "major cost." While it is relatively straightforward (though perhaps not totally uncontroversial) to assess whether a cost is minimal, or comparatively lower than the harm that is to be prevented as a result of its incursion, the notion of a major cost seems considerably less precise. How, then, should this concept be understood? I propose that we understand the concept of a major cost as one whose incursion would significantly impact one's ability to pursue important life projects or other significant aspects of their conceptions of the good. This understanding is non-arbitrary, as it follows directly from the demandingness objection raised against the strong interpretation of premise (1). If a strong interpretation is implausible because consistently acting upon it would make it impossible to live an autonomous life characterized by valuable projects and relationships, then the ability to live such a life must factor into the moderate interpretation's understanding of what constitutes an excessive cost. Of course, there may still be uncertainty about whether the costs in a particular instance are in fact major, and hence not morally required in the service of preventing serious harm to others, though the test for determining this is at least clear: we must assess whether incurring those particular costs would prevent one from pursuing an important life project or some other salient aspect of one's conception of the good.

A moderate interpretation is, therefore, the most plausible interpretation of premise (1) and is seemingly the only interpretation available to proponents of the rescue-based argument. But is it true as a general principle? It certainly seems to be true in the cases that are typically appealed to in order to generate a duty to rescue in the first place, such as *Shallow Pond*. In these types of cases, one's intervention is both *necessary* and *sufficient* to save the victim. However, there are other cases in which one's intervention is merely sufficient to save the victim due to the presence of other potential rescuers, and in these cases the general truth of premise (1) is much less apparent. We can distinguish between two types of cases that have this structure:

a The first type includes cases in which there are multiple potential rescuers present who are *identically situated* with respect to the victim, e.g., a version of *Shallow Pond* in which there are multiple strangers present, each of whom has an equal capacity to rescue the child.
b The second type includes cases in which there are multiple potential rescuers present who are *non-identically situated* with respect to the victim due to differences in their personal or relational properties, e.g., a version of *Shallow Pond* in which, of the multiple persons present, each has a different set of capacities and/or connections to the child that may be relevant to the distribution of responsibilities (e.g., one person pushed the child into the pond, one person is a close family member, one person is a professional lifeguard, and so on).

In neither of these cases does the mere capacity to prevent harm at less than a major cost *necessarily* generate a moral duty to do so. In cases that resemble (a), the duty to prevent harm is arguably held by the collective and the costs of discharging that duty ought to be distributed fairly between the individuals that compose it, either by dividing the costs according to some fair distributive principle or, if this is not possible due to the nature of the costs involved, by utilizing a fair procedure to determine who must totally incur them. In cases that resemble (b), considerations other than mere capacity may be equally or more relevant for determining who is obligated to render assistance to a victim, such as being morally responsible for the victim's predicament, being in an existing special or associative relationship with the victim, or being in possession of special capacities that enable one to render assistance more effectively or at a comparatively lower cost.[19] Of course, in both types of case, one may end up with a secondary obligation to come to a victim's assistance in the event of non-compliance by the primary duty-bearer(s) (if we accept there are duties to "take up the slack" left by others' non-compliance),[20] though notice that this would ground a legitimate complaint about having to incur the costs of assistance and would arguably generate duties of compensation in those who have failed to discharge their duties. These important features of the moral situation are not accounted for by the moderate interpretation of premise (1).

If these remarks are correct, they show that even a moderate interpretation of premise (1) is not generally true: the mere capacity to prevent harm at less than major cost does not necessarily generate a moral duty to do so in cases where one's intervention is merely sufficient to prevent serious harm. This is problematic for the rescue-based argument, as the orphan crisis is clearly a case in which one's intervention is merely sufficient to prevent serious harm. Positing a version of premise (1) that is generally true would require specifying the conditions under which the mere capacity to prevent serious harm generates a moral duty to do so, though, within the context of the rescue-based argument, this would at best generate a *conditional* duty to adopt and hence would not support the view that procreation is presumptively impermissible when there are existing children in need of parental care.

6.4 Can a Moderate Duty to Rescue Generate a Duty to Adopt?

Even if we accept that the mere capacity to prevent serious harm at less than major cost generates a moral duty to do so—either in general or in cases that meet an additional set of conditions—we still might reject the idea that this principle applies to the orphan crisis on the grounds that adopting a child typically involves a variety of costs, risks, and burdens that exceed a major cost threshold. There are in fact two different versions of this objection to premise (3), both of which can be understood as case-specific versions of the demandingness objection: one that focuses on the costs associated with forgoing biological parenthood, which apply to all cases of adopting rather than procreating as a means of satisfying one's interest in parenting, and one that focuses on the significant financial, emotional, and agency-related costs of adoption, many of which apply only contingently given existing social arrangements. I will argue in what follows that while the first version fails, the second often succeeds, such that in the context of existing social circumstances, there is rarely an individual duty to adopt based on a more general duty to rescue or assist those in need.

6.4.1 *The Cost of Forgoing Biological Parenthood*

To set the stage for the first objection, consider Rulli's *Safe Haven* case:

> *Safe Haven*: Imagine that you have just decided to become a parent. You learn that the local fire station, a safe haven, has received a newborn baby in need of a family. You are aware of the research showing that early infant adoptions pose little risk regarding the infant's psychological health and potential for emotional attachment. The urgent need for

> placement and the lack of administrative costs allow for the baby's adoption with few additional hurdles or financial burdens. You can adopt this child, who will otherwise face a life of uncertainty in various institutions or foster homes. Or you can decline and bring a new child into the world instead.[21]

In this case, it seems like you have a significant moral reason to adopt the baby from the fire station instead of bringing a new child into existence, for in so doing, you will be satisfying your own desire to become a parent *as well as* the baby's need to be parented. Moreover, the socially contingent costs that might have dissuaded you in other circumstances are not present: there are no costly adoption fees, no lengthy wait times, and the baby does not present any behavioral or health-related challenges that would significantly impact the parenting experience. Despite these considerations weighing in favor of adoption, however, many people still believe you are morally permitted to create a new child if doing so is sufficiently important to you. On this view, adopting the baby from the fire station is morally supererogatory: it is praiseworthy from a moral point of view, but not morally required in light of your strong preference for a biological child.

Can a strong preference for a biological child ground an exemption to the duty to adopt? If the argument in the previous section is correct, then this will depend on whether the costs associated with forgoing biological parenthood are appropriately considered to be major costs. I suggested earlier that the test for assessing whether a particular set of costs is major, and hence not morally required in the service of preventing serious harm to others, involves determining whether the incursion of those costs would prevent an agent from pursuing an important life project or some other significant aspect of their conception of the good. At first glance, it seems like preferences for biological children might very well pass this test, for many people sincerely regard creating and raising biologically related children as a central component of their conception of the good. In fact, the preference for biological children is so strong among some prospective parents that adoption fails to even register among their childrearing preferences. In other words, the preference ranking for some prospective parents is not (1) create a child, (2) adopt a child, or (3) have no children, but rather (1) create a child or (2) have no children.

Both Rulli and Friedrich concede that agents can be exempted from the duty to rescue when discharging that duty would conflict in certain ways with project-related interests—recall that this is precisely why they believe the duty to adopt cannot apply to non-parents. However, in cases where grave harm is at stake, they argue that exemption-grounding projects must also meet additional conditions. For Rulli, such projects must be *objectively valuable*,[22] whereas, for Friedrich, they must not be based upon *false beliefs*.[23] These additional conditions seem plausible when applied to standard rescue cases. Imagine, for example, that I refuse to rescue the child in *Shallow Pond* because it conflicts with my "project" of achieving a high score on my smartphone video game, or because I am under the demonstrably false belief that the water in the pond is electrified. Even though both of these excuses appeal to various aspects of my conception of the good, neither seems weighty enough to justify the non-performance of the duty to rescue given its triviality and/or basis in false beliefs.[24]

Rulli and Friedrich argue that similar reasoning applies with respect to the multitude of reasons that are commonly offered in favor of preferring biological over adopted children, which (a) are too trivial to justify the non-performance of the duty to adopt, (b) are based on false beliefs about adoption and the value of biological connectedness, or (c) fail to mark a morally relevant distinction between procreation and adoption. Having defended an account of the interest in procreation that is entirely derivative of the interest in parenting in Chapter 1, I am mostly persuaded by these arguments and hence do not believe that the first objection to premise (3) succeeds. Nevertheless, given the sincerity with which many prospective parents hold preferences for biological children, and the central role these preferences play in justifying their decisions to procreate rather than adopt, it is worth considering some of the major reasons offered for these preferences in greater detail. This will help to demonstrate why a successful objection to premise (3) of the rescue-based argument must appeal to other types of costs.

When confronted with the case of *Safe Haven*, we can distinguish between two categories of reasons that might be offered in favor of procreating rather than adopting the child from the fire station: those that have to do with the *process* of procreation and those that have to do with its *product*. One type of reason that falls into the former category appeals to the experience of *creating a new human life*. There are a number of reasons why creating a new human life might be considered valuable. Matthew Liao, for example, has likened the creation of a new human life to the creation of a beautiful work of art, as in both cases individuals are exercising their agency and autonomy to produce something with inherent value.[25] Carson Strong, by contrast, has argued that creating a new human life is instrumentally valuable because "intentionally having children can be an affirmation of a couple's mutual love and acceptance of one another" and can further provide "an anticipation that the bond between the couple will grow stronger because of common children to whom each has a biological relationship."[26] To the extent that these goods are both valuable and unique to procreation, they might provide a basis for preferring procreation over adoption.

However, it is not difficult to see that even if these goods are valuable, neither of them is unique to nor necessarily involved in procreation. First, if the value of creativity consists in exercising one's agency and autonomy in a particular way, then procreation actually seems like a poor candidate for an act that is creative in a valuable sense. While a set of individuals may autonomously engage in reproductive sex, what happens afterward is almost entirely determined by biological processes that are beyond their control and is therefore a poor analog for a deliberately creative act like artistic expression. There may of course be creative value in shaping a child's identity by parenting them once they are born,[27] though this form of creativity fails to mark a distinction between procreation and adoption, as it is available to biological and adoptive parents in equal measure.

Second, while raising a child in the context of a shared family life may well serve to affirm a couple's mutual love and respect for one another, it is less plausible to think that this type of affirmation is brought about by procreation alone. If this were the case, it would imply that a couple who procreates on the basis of a one-night stand and subsequently puts their child up for adoption can expect their bond to grow stronger in light of the fact that there happens to be a child in the world to whom they are both biologically related. This implication is clearly false, however, suggesting that if having a child serves to affirm a couple's mutual love and affection for one another, it is only to the extent that they actually share in the task of raising that child—a task that can be performed with respect to both biological *and* adopted children.[28]

A second process-based reason for preferring procreation to adoption appeals to the *experience of pregnancy*: only by procreating is a woman able to have the unique experience of gestating and giving birth to another human being. The desire to experience pregnancy is often strong among many prospective mothers, and the experience itself is often thought to have both instrumental and intrinsic value. For instance, some commentators have argued that pregnancy is instrumentally valuable insofar as it prepares women for motherhood and contributes to the development of an intimate maternal-infant bond,[29] while others have argued that it is intrinsically valuable insofar as the experience itself contributes in some way to the identity and self-conception of the women who pursue it.[30] Given that adoption does not involve experiencing pregnancy, a desire to have that experience might be thought to support a legitimate preference for procreation over adoption.

According to Rulli, the desire to experience pregnancy is the *only* legitimate reason that prospective parents can advance in favor of their preference to have a biological child, for, unlike the other reasons that are typically advanced in favor of that preference, "the interest in pregnancy may arise to the level of a project."[31] The desire to experience pregnancy offers a stronger basis than the desire to create a life for grounding an exemption to the duty to adopt, though one might still question whether it merits being granted this exceptional status. First, the apparent instrumental value of pregnancy seems overstated. While it is commonly assumed that mothers have closer emotional connections to their biological as opposed to adoptive children, I noted in Chapter 3 that numerous empirical studies have suggested that mother-infant attachment is usually as strong in the case of adoption as it is in the case of biological procreation, particularly when children are adopted early in their infancy.[32] Moreover, it is also unclear whether pregnancy necessarily gives biological mothers a competitive edge in the preparation for parenthood. While pregnancy may help to prepare mothers for many of the changes brought about by parenthood, so may the rigorous process that adoptive parents often undergo before adopting a child, which typically involves prolonged waiting periods, home visits, and parenting classes. Finally, there are also certain respects in which adoption may have instrumental advantages over pregnancy. While adoptive mothers do not experience the benefits of pregnancy, they also do not face its associated risks, including the risk of miscarriage, pre-eclampsia, or prenatal and post-partum depression.[33]

Second, even if pregnancy has intrinsic value insofar as it is a unique and non-substitutable experience, one may question whether it is sufficiently important to justify an exemption to the duty to adopt. In the *Safe Haven* scenario outlined above, citing the desire to experience pregnancy as the basis for an exemption might seem to imply that the cost associated with forgoing that experience is morally on par with the cost to the child of not being adopted. However, given the cost to children of going without parental care, some may find this comparison to be implausible, suggesting that the mere desire to experience pregnancy cannot exempt women from the duty to adopt.

In addition to reasons pertaining to the process of procreation, a second category of reasons for preferring procreation to adoption appeals to its product, namely, a child who is biologically related to oneself. One set of reasons that falls into this second category appeals to the apparent value of *physical or psychological resemblance*. It is common for parents to react in joy or amusement when various physical traits are reproduced in their offspring—for example, discovering that a baby has its father's eyes or its mother's hair is usually cause for wonder and celebration, and it is an experience that is unique to biological parenthood. Similarly, there may also be value in seeing one's offspring share in what Brenda Almond has termed one's *Weltanschauung*, or the various "attitudes, appraisals, interests, tendencies, [and] common qualities of character" that make up one's "characteristic way of looking at the world."[34] To the extent that these resemblances are both valuable and biologically determined, they might provide an additional basis on which to prefer procreation to adoption.

However, there is good reason to be skeptical of both the value of these resemblances and the extent to which they are genetically determined.

Consider first the value of physical resemblance. Even if procreation is more likely to result in a child that physically resembles its parents, it is difficult to see why the desire to achieve that kind of resemblance should be assigned any normative weight. At best, it reflects a trivial curiosity that is not weighty enough to justify the non-performance of a duty; at worst, it reflects an objectionable form of narcissism that ought to be discouraged rather than legitimized. It is also worth noting here that a desire for physical similarity might lead one to counterintuitive conclusions when generalized to other cases. For example, if there is genuine value in creating a child who is physically similar to oneself, then this would imply that cloning a child would be preferable to creating one through procreation.

The desire to achieve psychological similarity might carry greater normative weight insofar as it contributes to the quality of the parent-child relationship, though the problem here is that the relevant psychological features are not genetically determined, but are rather the product of complex interactions between a range of genetic and environmental factors.[35] As Neil Levy and Mianna Lotz have argued, to assume that one's progeny will share one's *Weltanschauung* simply in virtue of sharing one's genes is to engage in a crude form of genetic determinism and to underplay the significant role of environmental factors in shaping personality.[36]

A final common reason that prospective parents might advance in favor of their preference for a biological child is the desire to achieve a form of *immortality* by having their genetic material live on into the future. The desire the achieve immortality is sometimes cited as an unconscious desire motivating the decision to procreate—as John Robertson claims, "in connecting us with nature and future generations, reproduction gives solace in the face of death"[37]—though cases of posthumous reproduction have given it a more concrete expression. For example, it is not uncommon for terminally ill men to make their sperm available for their female partners' use following their deaths, and there have even been cases of post-mortem sperm retrieval in cases where men have died accidentally before being able to act on an expressed desire to procreate. While less common, there have also been cases in which terminally ill women have made their ova available to their male partners for fertilization and gestation by surrogate following their deaths, attesting to the fact that the desire to pass on one's genetic material can be very strong among many prospective parents.[38]

Can the desire to achieve immortality ground an exemption to the duty to adopt? There are a number of reasons to be skeptical that it can. First, procreation is no guarantee of genetic immortality. In order for one's genetic material to live on into the future, it must be the case that one's children procreate as well, though there is never any certainty that they will choose or be able to do so—in this sense, procreation only guarantees that one's genetic material will live on for *one* generation. Moreover, even in cases where one's genetic material *does* survive into the future, it diminishes in significance with each subsequent generation. While one's progeny will share 50% of one's genetic material, one's grandchildren will only share 25%, one's great-grandchildren 12.5%, and so on, until the proportion of one's genetic material in each descendant is inconsequential.[39]

Second, and more to the point, it is also unclear that genetic self-extension is the most meaningful way to achieve the relevant form of immortality. Imagine that one's choice was between (a) creating a genetically related child who grew to reject one's culture, values, and traditions and (b) adopting a non-genetically related child who grew to embrace them. If one's decision to become a parent was in part motivated by a desire to achieve a lasting form of immortality in the sense of influencing future generations, it seems like one would be more successful choosing (b) over (a).

If the preceding arguments are sound, then many of the reasons commonly advanced in favor of preferring biological over adoptive children do not seem weighty enough to justify procreation over adoption: some are based on false assumptions about the value of biological connectedness, others seem too trivial to justify the non-performance of the duty, and others fail to mark a meaningful distinction between procreation and adoption given that the purportedly valuable features of procreation often have close analogs in cases of adoption. Notice, however, that even if one rejects these arguments and insists that reasons for preferring biological children *are* morally significant, they may not ground an exemption to the duty to adopt in all cases. Given that each of the aforementioned interests can be satisfied by the creation of just *one* biological child, appealing to any of them may only ground an exemption to the duty to adopt *once*. The duty to adopt may still apply with respect to any subsequent children that one wanted to parent, as one's preference for a biological child would then have been satisfied.

6.4.2 *The Financial, Emotional, and Agency-Related Costs of Adoption*

Of course, the preference for biological children is not the only consideration dissuading prospective parents from adoption. In justifying their decision to procreate rather than adopt, many prospective parents are just as likely to appeal to the significant financial, emotional, and agency-related costs of adoption. It is no secret that, in many societies, adopting a child remains very expensive. In the United States, for example, the average cost of adopting a child is between $10,000 and $15,000, though it is not uncommon for costs to exceed $40,000 depending on the type of adoption (e.g., domestic or international), the type of agency that is facilitating it (e.g., public or private), as well as the age and health status of the prospective adoptee. Some jurisdictions offer income-proportionate tax credits or adoption subsidies to defray some of these costs, though significant credits are often reserved for adoptions involving children with special needs, who also tend to be more expensive to raise.[40]

The process of adopting a child also involves a number of legal and logistical hurdles that procreative parents do not typically have to overcome. For example, adoptive parents must typically undergo an invasive parental licensing process, which usually involves a comprehensive home study, a requirement to attend parenting classes, and close scrutiny of their personal and family background.[41] Many adoptive parents must also endure longer wait times than procreative parents before gaining custody of their child. While a successful pregnancy can be achieved in as little as nine months, a finalized adoption can sometimes take years to complete, during which time adoptive parents can be left in a state of uncertainty about whether their adoption will in fact proceed. These hurdles combine to make the process of adoption a frustrating and emotionally taxing experience, or as Carolyn McLeod and Andrew Botterell have put it, one that is "not for the faint of heart."[42]

Finally, adoptive parents may also face a unique set of challenges in raising the children they successfully adopt into their families. In contrast to Rulli's *Safe Haven* case, the majority of children available for adoption are not healthy newborns, but are rather children over the age of five who may have spent significant amounts of time in institutional or foster care settings.[43] This is important for at least two reasons. First, children who have spent significant time in institutional care are more likely to present a range of developmental and behavioral problems that create additional challenges for adoptive parents and that qualitatively impact the parenting experience. Some prospective parents may be prepared to undertake these challenges, though others may find them too demanding, preferring to forgo parenthood altogether. Second, the fact that the majority of children available for adoption are over the age of five means that, by adopting rather than procreating as a means of fulfilling their interest in parenting, many prospective parents must forgo the experience of raising a child from its infancy. This may plausibly register as a major cost insofar as the experience of raising a child from its infancy is a non-trivial and non-substitutable element of one's conception of the good.

The significant financial, emotional, and agency-related costs of adoption form the basis of a second type of objection to premise (3). Even if we do not assign any weight to individual preferences for biological children, it still might be the case that the various other costs associated with adoption in existing social circumstances are too high to be morally required. To get a sense of the intuitive plausibility of this objection, consider a modified version of Rulli's *Safe Haven* case that accounts for some of the socially contingent costs of adoption:

> *International Adoption*: Imagine that you have just decided to become a parent. You learn of an orphanage in an overseas country that has several children available for adoption. The ages of these children range between 5 and 14 and their physical and psychological health status is uncertain. After conducting some research, you learn that adopting one of these children may take up to two years, cost between $25,000 and $40,000 in fees and travel costs, and will involve navigating a complicated international adoption bureaucracy with no guarantee of success. You can initiate the process of adopting one of these children, who will otherwise face a life of uncertainty in various institutions or foster homes, or you can decline and bring a new child into the world instead.

It is far less certain whether you have a duty to adopt in *International Adoption*. Even if you attribute no significance to biological parenthood, there are many other reasons you might appeal to in order to justify procreation over adoption, including the significant financial implications of pursuing adoption, the emotional burden of the administrative process, or your strong desire to experience parenting a child from its infancy. On this view, while you may have a duty to adopt in idealized cases like *Safe Haven*, you do not necessarily have such a duty in real-world contexts where the costs of adoption are often considerably higher and more multifaceted.

Is this second objection to premise (3) sound? Some theorists have expressed their doubts.[44] Friedrich, for example, has argued that objections to the duty to adopt based on financial costs are often overstated, as they tend to ignore the costs associated with procreation that are avoided through adoption. For instance, he notes that the "medical costs of pregnancy, childbirth, and infant care are often significant and the costs of raising a child through its first year or two of life (including opportunity costs of missed wage income) are also substantial."[45] He notes further that these costs can expand considerably in cases where prospective parents require access to assisted reproductive technologies (ARTs) like *in vitro* fertilization (IVF). In the United States, for example, where IVF treatment is rarely subsidized, the cost per live birth (which includes the cost of unsuccessful cycles) can easily exceed $45,000, particularly for women over the age of 40 who may be required to undergo multiple cycles.[46] The case of IVF is particularly significant in this context, for it reveals not only that procreation can involve financial costs that are comparable to those associated with adoption but also that there are many prospective parents who are willing to incur them.

We might add to Friedrich's point about financial costs the fact that procreation can also involve a host of additional burdens and frustrations that are not dissimilar to those faced by adoptive parents. For example, while the process of adoption can involve false starts and excessive wait times, so too can the process of procreation: it can sometimes take couples months or even years to conceive a child, and many must endure the pain and worry associated with miscarriage and other pregnancy-related complications. Moreover, while some adoptive children may pose a higher risk of developmental problems due to post-institutional trauma, procreation may also result in children with congenital disabilities or other health-related problems that present significant challenges for their biological parents. When we take into account the burdens and risks associated with procreation, we begin to see that it is parenthood in general that is a demanding enterprise and that the differences between parenting via procreation and parenting via adoption are sometimes differences of degree rather than type.

These are valid points we must consider when assessing the overall costs of adoption. However, as I argued previously, the true test for whether these costs can be appealed to as an objection to premise (3) is whether they exceed a major cost threshold by preventing an agent from pursuing an important life project or other significant aspect of their conception of the good. There are at least two ways in which they might. First, the financial and emotional costs of adoption may be so burdensome for some prospective parents that incurring those costs would preclude the pursuit of other significant aspects of their conceptions of the good. Of course, whether this is true will depend on the costs associated with adoption and the financial and emotional capacities of prospective parents, both of which can vary significantly by context, though for some prospective parents, they will plausibly ground an exemption to the duty to adopt. Second, many and perhaps most prospective parents do have strong preferences for raising *infants* from the earliest stages of their lives, though the majority of adoptable children are older children whose infancy is already behind them. The preference for experiencing parenthood throughout a child's infancy is non-trivial and arguably non-substitutable: it is a unique period of dependency and vulnerability characterized by the formation of strong attachments, the witnessing of important developmental milestones, and the development of parental capacities and dispositions that contribute in significant ways to parents' well-being, including the special qualities of care, attention, and keen sensitivity to need that accompanies caring for a vulnerable infant. In forgoing this experience by adopting an older child rather than procreating, many prospective parents would be sacrificing an important component of their conception of the good.

If this second objection to premise (3) is sound, it reveals that unless we endorse a strong interpretation of premise (1), according to which we are morally required to prevent serious harm to others to the point of marginal utility, the rescue-based argument will most reliably generate a duty to adopt in a limited range of cases, that is, cases in which prospective parents are able to adopt a healthy infant without incurring excessive financial, emotional, or agency-related costs in the process. This is problematic for at least two reasons. First, as Liao has pointed out, such a duty appears to be of little significance given that there is no shortage of parents who are already willing to adopt healthy infants.[47] Data from the Population Division of the UN Department of Social and Economic Affairs suggests that, since the 1970s, there has been a widening gap between the number of families wishing to adopt and the number of babies waiting to be adopted, particularly in the context of domestic adoptions. In the United States, for example, it is estimated that there are at least three prospective families for every domestic baby available for adoption, with even wider gaps found in other countries for which there is available data, including Italy, France, and Singapore.[48] In cases like these, where the children to whom a duty to adopt applies are likely to be adopted anyway by willing volunteers, the duty itself disappears and there are no longer any adoption-related objections to procreation.

The picture becomes more complicated when we include the possibility of international adoption, for while there are likely to be more prospective adoptees relative to the number of families willing to adopt, the costs are also much higher and hence affordable to fewer prospective parents. There are also additional moral hazards in the context of international adoption that complicate matters further given significant financial incentives to service Western demand for adoptable children. For example, Kathryn Joyce has documented a troubling trend of children in developing countries being separated from their parents and funneled illegitimately into the adoption system as a result of child trafficking or temporary relinquishment due to poverty.[49] Preventing serious harm to these children would not entail adopting them, but rather equipping their parents with the resources to adequately support them within their existing family arrangements. Despite these complications, however, the general point remains and can simply be reframed conditionally: the rescue-based argument will most reliably generate a duty to adopt in cases where (a) one is able to adopt a healthy and legitimately adoptable infant without incurring excessive financial, emotional, or agency-related costs in the process, and (b) this infant is not likely to be adopted anyway by a willing volunteer.

The second problem with the limited applicability of a rescue-based duty to adopt is that it will simply not apply to most orphaned children, the majority of whom are estimated to be over the age of five, and many of whom are likely to have special needs related to post-institutional trauma that may fundamentally impact the parenting experience in ways that are morally significant.[50] This seems unsatisfactory. After all, it was the great magnitude of the global orphan crisis that prompted the development of the rescue-based argument in the first place, though the most defensible interpretation of that argument seems unable to generate duties with respect to the majority of those children.

6.5 Collective Obligations to Facilitate Adoption: The Case of ART

In this chapter, I have presented two arguments against the rescue-based argument for the individual duty to adopt. The first argument challenged premise (1) on the grounds that the mere capacity to prevent serious harm at less than major cost does not necessarily generate a moral duty to do so in cases where one's intervention is merely sufficient to prevent serious harm. If there are others available who (a) have sufficient capacity to prevent serious harm to a victim and/or (b) have additional moral reasons to come to a victim's assistance in virtue of some relevant personal or relational property, then one may only be required to come to the victim's assistance if one is allocated that responsibility through a fair distributive procedure or if one is otherwise required to do so in the event of non-compliance by the primary duty-bearer(s). The second argument challenged premise (3) by pointing to the significant costs that are associated with adopting children in current, non-ideal conditions, with a special emphasis placed on the costs associated with forgoing the opportunity to parent a child from its infancy. I suggested that these costs will often exceed a major cost threshold and hence would not be morally required by a moderate interpretation of premise (1). This significantly limits the constituency of children to whom a rescue-based duty to adopt applies, with many of those children likely to be adopted anyway by willing volunteers.

What should we conclude on the basis of these arguments? One conclusion we can draw is that the rescue-based argument fails to establish that procreation is wrong in cases where there are already existing children in need of parental care. A token instance of procreation *may* be impermissible on the terms of the rescue-based argument, but this will depend on a number of variables, including the presence and characteristics of other potential adopters, the preferences and capacities of the procreators themselves, and the costs that are associated with adopting a child in that particular circumstance. It will sometimes turn out that these variables support a duty to adopt rather than procreate, though it will also and perhaps more often turn out that they support an exemption to that duty.

A second conclusion we can draw is that, despite its noble intentions, the rescue-based argument actually seems like a poor foundation for our moral obligations to orphaned children. The most plausible interpretation of that argument restricts the duty to adopt to prospective parents who are able to adopt orphaned children at less than major cost to themselves, though this accounts for only a portion of the total number of children in need of parental care, leaving the remainder outside the scope of the duty. It also appears to place a disproportionate burden on prospective parents for resolving the orphan crisis, understating both the costs to prospective parents of doing so and the capacity of non-parents to shoulder some of the costs associated with adoption.

Thus, advocates of adoption ought to look beyond the duty to rescue to ground our moral obligations to orphaned children. As Rulli herself has suggested, one promising way forward may be to shift our focus from individual duties to adoption to collective obligations to facilitate adoption and to otherwise improve the lives of orphaned children.[51] I would like to conclude my discussion by examining one interesting policy proposal that might function to socialize the significant costs that are associated with adoption while incentivizing at least one subset of prospective parents to consider adoption over procreation: redirecting subsidies away from supporting fertility treatment to support adoptive parents.

I noted earlier that in countries like the United States, where IVF treatment is rarely state-funded, the cost of achieving a live birth can often be substantial, particularly for older women who may be required to undergo multiple cycles. However, this is not the case in all countries. In the United Kingdom, for example, the National Health Service (NHS) funds up to three cycles of IVF for women under the age of 40, and one cycle for women aged 40–42, while in Israel IVF is entirely state funded up until the birth of two children for women under the age of 45, regardless of the number of cycles it takes to achieve that outcome. Many other countries offer at least partial subsidization of IVF treatment, including Austria, Spain, Sweden, the Netherlands, Finland, and Canada, to name just a few.[52]

Is it justifiable for states to invest significant public resources in funding expensive fertility treatments when there are children—some of whom reside *within* their own borders—in need of parental care? Jurgen De Wispelaere and Daniel Weinstock have argued otherwise.[53] Noting an asymmetry in need fulfillment between the case of adoption and the case of ART, De Wispelaere and Weinstock have argued that states have a significant reason to subsidize adoption over ART, for while subsidizing ART allows them to satisfy the needs of only infertile prospective parents, subsidizing adoption allows them to satisfy the needs of infertile prospective parents as well as the needs of orphaned children, reflecting a more just and efficient use of public resources. As they explain,

> Adoptive children have an interest in restricting the availability of competing reproductive technologies as this affects their chances of finding an adoptive family. Children who owe their existence to ART, on the other hand, have no such interest, for if ART were not made available to prospective parents, these children would simply not exist. In other words, where 'adoptive children' can be harmed by promoting ART, 'ART children' cannot be similarly harmed by promoting adoption. This asymmetrical dimension of ART versus adoption gives the state a legitimate reason to promote adoption over ART, a reason that is furthermore proportional to the extent that ART directly affects adoptive children's interests. At the same time, prospective parents have no valid reason for complaint, for their fundamental right to parent is satisfied through adoption even when they are denied access to parent their own biological child through ART.[54]

Crucial to De Wispelaere and Weinstock's argument is the assumption that the valuable aspects of parenting can be achieved just as well in adoptive parent-child relationships as they can in biological parent-child relationships. If there was something distinctively valuable about the quality of the relationship that was achievable between parents and their biological children, then infertile prospective parents may have a legitimate complaint against a funding scheme that promoted adoption over ART, as parenthood via adoption would be in a relevant sense second-best. If, however, the valuable aspects of parenting can be achieved just as well by parenting adopted children as they can by parenting biological children, then infertile prospective parents have no such complaint, as their fundamental right to parent can be still be satisfied through adoption. Any remaining preference for biological children amounts to something like an "expensive taste," or a discretionary interest that states are under no obligation to accommodate.

Having rejected the notion that parenthood via procreation is valuable in a way that parenthood via adoption is not, I am inclined to find De Wispelaere and Weinstock's argument persuasive. However, before endorsing the view that states ought to redirect resources away from ART and toward adoption, there are a number of possible objections to take into consideration. First, one might object that a number of empirical realities ultimately undermine the extent to which ART can be substituted by adoption. The argument for funding adoption over ART seems to assume that the two are more or less interchangeable, though for reasons outlined in the previous section, this is not necessarily the case. For instance, while ART provides prospective parents the possibility of parenting a healthy newborn child, adoption may only provide prospective parents the opportunity to parent a child who is over the age of five, and who may or may not exhibit a range of development problems as a result of long-term institutional care. Similarly, while prospective parents can often utilize ART with minimal scrutiny of their personal and family background, adopting a child in most jurisdictions still involves an onerous and invasive parental licensing process. In this sense, even if we agree that there are no significant differences between parenthood via procreation and parenthood via adoption in principle, it may still be the case that there are significant differences between them in practice.

De Wispelaere and Weinstock are sensitive to this type of objection and temper their view accordingly, concluding that "access to adoption by and large remains too complex and burdensome for the time being to be able to justify significant restrictions on the availability of ART for those prospective parents who cannot conceive naturally."[55] However, while it may be true that adoption is currently too burdensome to constitute a perfect substitute for procreation, it is not clear that this necessarily weighs against the proposal to subsidize adoption over ART. When evaluating these types of proposals, it is important not to lose sight of the fact that the current orphan crisis is partially a function of the various laws, policies, and social norms that govern adoption. One of the reasons that orphaned children remain so long in institutional care, and subsequently develop a range of developmental and behavioral problems that make them more difficult to raise, is that there are laws and policies in place that make adoption excessively costly—thus, citing the difficulty of raising adopted children as a reason against amending those laws and policies seems like a potentially question-begging response. If we took measures to make adoption a less costly and hence more attractive option—e.g., by redirecting public resources from ART to adoption—then orphaned children may ultimately spend less time in institutional care, and thereby avoid developing the kind of cognitive and behavioral problems stemming from post-institutional trauma that dissuade prospective parents from adopting them in the first place.

Similarly, while it is true that adoptive parents currently face a range of legal and bureaucratic hurdles that ART parents do not, these hurdles are again the product of existing laws and policies and do not therefore count as a consideration against the proposal to subsidize adoption over ART. As De Wispelaere and Weinstock point out, rather than maintaining the status quo on parental screening in adoption versus ART, "the state could equalize the current disparity between very strict adoptive screening and comparatively lax screening for ART,"[56] either by relaxing the strict screening standards that currently apply to adoptive parents, introducing more stringent screening standards for accessing ART, or both.

However, a second and arguably more troubling objection to the proposal to promote adoption over ART is that it is discriminatory, for it seems to place a disproportionate burden on infertile individuals to meet the needs of orphaned children (or, perhaps more accurately, infertile individuals who cannot afford to pay the cost of their own fertility treatment). As Eduardo Rivera-López has notably argued, this type of singling-out seems arbitrary: to the extent that infertility is an unchosen physical attribute, singling out infertile individuals as special duty bearers toward orphaned children makes just as much sense as singling out individuals with a height of over 1.70 meters, or individuals with blue eyes.[57]

Is this objection sound? Does privileging adoption over ART place an unfair burden on infertile individuals to meet the needs of orphaned children? As a first response to this objection, it is important to note that the argument for subsidizing adoption over ART does not assume that we should *only* promote adoption among infertile individuals. On the contrary, it may be pursued alongside other policies designed to promote adoption as a valuable childrearing option. For instance, in addition to promoting adoption over ART, states might work to challenge bionormativity through public education (much in the way that heteronormativity is currently being challenged in many public school curriculums), reduce bureaucratic barriers to adoption that currently work to disincentivize prospective adoptive parents, and socialize the costs of adoption generally through taxation, similar to the way that they fund other important social programs, and work to reinforce the idea that adoptive parent-child relationships can have a unique type of value that biological parent-child relationships lack—namely, the value associated with helping an existing child in need.[58]

However, even if, despite these measures, subsidizing adoption over ART *did* create a form of disparity between fertile and infertile individuals, this might be an acceptable outcome given the available alternatives. Consider, for instance, the two obvious ways in which the state might achieve the relevant form of parity between fertile and infertile individuals. First, the state might achieve parity by attaching costs to natural procreation that are equivalent to the costs associated with unfunded ART, e.g., by implementing a tax on natural procreation when there are already existing children in need of parental care. This way, fertile prospective parents would have the same economic incentive to adopt rather than procreate that infertile prospective parents have under a scheme of unfunded ART. Alternatively, the state could fully subsidize ART so that infertile individuals have the same opportunity to procreate as fertile individuals. This way, both fertile and infertile individuals would be free to procreate if they so choose, or adopt children on a fully voluntary basis.

Both of these options have their own disadvantages. While the second option works to ameliorate the disadvantage faced by infertile individuals, and may therefore be attractive to those who see infertility as a disability or form of brute bad luck, it is worse than the status quo from the perspective of orphaned children, who may be even less likely to be adopted into families. Similarly, while the first option may be better from the perspective of orphaned children, it also carries a potentially negative side-effect, for by heavily taxing those who procreate naturally when there are existing children in need of parental care, the state may be indirectly disadvantaging their biological children.[59] Given the available alternatives, and taking into account the needs of orphaned children, subsidizing adoption over ART does not seem morally objectionable even if it does create a form of disparity between infertile and fertile individuals. While it would be impermissible for states to prohibit ART outright, or actively preclude infertile individuals from accessing it, there does not seem to be anything particularly objectionable about the state biasing its funding decisions in favor of the interests of orphaned children.

Notes

1 UNAIDS, UNICEF, and USAID. *Children on the Brink 2004: A Joint Report of New Orphan Estimates and a Framework for Action* (New York: United Nations Children's Fund, 2004), 4. The precise number of orphans is notoriously difficult to calculate not only due to factors like chronic underreporting and the hidden nature of child abandonment but also to the fact that the definition of "orphan" varies according to context. The magnitude of the global orphan crisis is sometimes grossly exaggerated by commentators who appeal to statistics based on a more expansive definition of an orphan, e.g., as a child who has lost one or more parents, of which there are currently over 132 million.

2 Elizabeth Bartholet. "International Adoption: The Human Rights Position," *Global Policy*, 1(1) (2010), pp. 91–100, 95.

3 Tina Rulli. "Preferring a Genetically-Related Child," *Journal of Moral Philosophy*, 13(6) (2016), pp. 669–698.

4 Rulli, "Preferring a Genetically-Related Child," 674.

5 For an overview of the empirical literature on the developmental consequences of institutional and foster care, see Elizabeth Bartholet. *Nobody's Children: Abuse and Neglect, Foster Drift, and the Adoption Alternative* (Boston: Beacon Press, 1999), chapters 3 and 4.

6 Daniel Friedrich. "A Duty to Adopt?" *Journal of Applied Philosophy*, 30(1) (2013), pp. 25–39; Jurgen de Wispeleare and Daniel Weinstock. "State Regulation and Assisted Reproduction: Balancing the Interests of Parents and Children" in François Baylis and Carolyn McLeod (eds.) *Family-Making: Contemporary Ethical Challenges* (Oxford: Oxford University Press, 2014) and "Privileging Adoption Over Sexual Reproduction? A State-Centered Perspective" in Sarah Hannan, Samantha Brennan, and Richard Vernon (eds.) *Permissible Progeny? The Morality of Procreation and Parenting* (New York: Oxford University Press, 2015); Thomas Søbirk Petersen. "The Claim from Adoption," *Bioethics*, 16(4) (2002), pp. 353–374; and Rulli, "Preferring a Genetically-Related Child," "The Duty to Adopt" (PhD Diss., Yale University, 2012), "The Unique Value of Adoption," in François Baylis and Carolyn McLeod (eds.) *Family-Making: Contemporary Ethical Challenges* (Oxford: Oxford University Press, 2014), and "The Ethics of Procreation and Adoption," *Philosophy Compass*, 11(6) (2016), pp. 305–316.

7 This is a version of Peter Singer's *Shallow Pond* case presented by Peter Unger in *Living High and Letting Die: Our Illusion of Innocence* (Oxford: Oxford University Press, 1996), 9. For Singer's original case, see his "Famine, Affluence, and Morality," *Philosophy and Public Affairs*, 1(3) (1972), pp. 229–243, 231.
8 Singer, "Famine, Affluence, and Morality," 231. Singer's original claim is that "if it is in our power to prevent something bad from happening, without thereby sacrificing anything of comparable moral importance, we ought, morally, to do it."
9 Friedrich, "A Duty to Adopt?" 25–26; Rulli, "The Duty to Adopt," chapter 2 and "Preferring a Genetically-Related Child," 672–678. For a more recent exploration of Friedrich and Rulli's rescue-based arguments, see also Travis N. Rieder. "Procreation, Adoption, and the Contours of Obligation." *Journal of Applied Philosophy*, 32 (3) (2015), pp. 293–309.
10 Rulli, "The Duty to Adopt," 71.
11 Rulli, "The Duty to Adopt," 55–63. See also Bartholet, *Nobody's Children*, chapters 3 and 4.
12 Rulli, "The Duty to Adopt," 62.
13 For a prominent defense of this view, see Liam Murphy, "The Demands of Beneficence," *Philosophy and Public Affairs*, 22(4) (1993), pp. 267–292.
14 Rulli, "Preferring a Genetically-Related Child," 8.
15 Singer, "Famine, Affluence, and Morality," 241.
16 See, for example, Samuel Scheffler. *The Rejection of Consequentialism* (Oxford: Clarendon Press, 1994); Bernard Williams. "A Critique of Utilitarianism" in J.J.C. Smart and Bernard Williams. *Utilitarianism: For and Against* (Cambridge: Cambridge University Press, 1973); and Susan Wolf. "Moral Saints," *Journal of Philosophy*, 79(8) (1982), pp. 419–439.
17 The anti-natal implications of a strong interpretation of the duty to rescue have been explored previously by Stuart Rachels. "The Immorality of Having Children," *Ethical Theory and Moral Practice*, 17(3) (2014), pp. 567–582, and David Benatar. "Famine, Affluence, and Procreation: Peter Singer and Anti-Natalism Lite," *Ethical Theory and Moral Practice*, 23(2) (2020), pp. 415–431.
18 As I argued in Chapter 5, presently existing people depend on subsequent generations to support the social and economic institutions they will depend on in old age, though they also may depend on them in a deeper sense to the extent that the value of our life projects is dependent on the assumption of a "collective afterlife" in which others will live on after we die. For a prominent defense of the latter view, see Samuel Scheffler. *Death and the Afterlife*. Ed. Niko Kolodny (Oxford: Oxford University Press, 2013).
19 For a discussion of the various grounds beyond capacity on which duties of assistance might be assigned, see David Miller's connection theory of remedial responsibility, presented in "Distributing Responsibilities," *Journal of Political Philosophy*, 9(4) (2001), pp. 453–471.
20 For a defense of these duties, see Zofia Stemplowska, "Doing More than One's Fair Share," *Critical Review of International Social and Political Philosophy*, 19(5) (2016), pp. 591–608.
21 Rulli, "Preferring a Genetically-Related Child," 669–670.
22 Rulli, "Preferring a Genetically-Related Child," 677–678.
23 Friedrich, "A Duty to Adopt?" 28–30.
24 It is worth noting here that there is an acknowledged risk for Rulli and Friedrich in insisting that exemption-grounding projects meet these additional conditions,

as it may undermine the view that non-parents are exempt from the duty to adopt based on their preference to remain childless. If this preference was based on false beliefs (e.g., that parenting a child is always a thankless chore with no compensating benefits) or trivial interests (e.g., to sleep in on the weekends), then it would not ground an exemption to the duty to adopt, opening the door once again to a moral obligation to become a parent.

25 S. Matthew Liao. *The Right to Be Loved* (Oxford: Oxford University Press, 2015), 155.

26 Carson Strong. "Cloning and Infertility," *Cambridge Quarterly of Healthcare Ethics*, 7 (1998), pp. 279–293, 282.

27 Colin Macleod in fact uses the term "creative self-extension" to describe this morally valuable feature of parenting. See "Parental Responsibilities in an Unjust World" in David Archard and David Benatar (eds.) *Procreation and Parenthood: The Ethics of Bearing and Rearing Children*. (Oxford: Oxford University Press, 2010), esp. at 142.

28 Also worth noting here is the fact that Strong's argument really only applies to heterosexual couples—it does not apply, for instance, to single individuals who procreate independently, or same-sex couples who do not share a biological connection with their child.

29 Anca Gheaus. "The Right to Parent One's Biological Baby," *Journal of Political Philosophy*, 20(4) (2012), pp. 432–455, esp. 446–451, and 453.

30 Rulli, "Preferring a Genetically-Related Child," 25–28. See also Amy Mullin. *Reconceiving Pregnancy and Childcare: Ethics, Experience, and Reproductive Labor* (New York: Cambridge University Press, 2005), esp. 27–34.

31 Rulli, "Preferring a Genetically-Related Child," 2014, 28.

32 See Chapter 3, Section 3.5.1.

33 Rulli, "Preferring a Genetically-Related Child," 27.

34 Brenda Almond. "Family Relationships and Reproductive Technology" in U. Narayan and J.J. Bartkowiak (eds.) *Having and Raising Children: Unconventional Families, Hard Choices, and the Social Good* (University Park, PA: Pennsylvania University Press, 1999), 104. See also Rivka Weinberg. *The Risk of a Lifetime: How, When, and Why Procreation May Be Permissible* (Oxford: Oxford University Press, 2015), pp. 41–44.

35 Neil Levy and Mianna Lotz. "Reproductive Cloning and a (kind of) Genetic Fallacy," *Bioethics*, 19(3) (2005), pp. 232–250, esp. 237–238.

36 Levy and Lotz. "Reproductive Cloning and a (kind of) Genetic Fallacy," 237–238.

37 For instance, in speculating on the significance of procreation, John Robertson suggests that "in connecting us with nature and future generations, reproduction gives solace in the face of death." See John A. Robertson. *Children of Choice: Freedom and the New Reproductive Technologies* (Princeton: Princeton University Press, 1994), 24.

38 For discussion of these and other cases of posthumous reproduction, see Carson Strong, Jeffrey R. Gingrich, and William H. Kutteh. "Ethics of Post-mortem Sperm Retrieval," *Human Reproduction*, 15(4) (2000), pp. 739–745.

39 Levy and Lotz, "Reproductive Cloning," 242.

40 Child Welfare Information Gateway. *Costs of Adopting: A Factsheet for Families*. (Washington, DC: U.S. Department of Health and Human Services, 2011), 6–7.

41 Andrew Botterell and Carolyn MacLeod. "'Not for the Faint of Heart': Assessing the Status Quo on Parental Licensing" in McLeod and Baylis, *Family Making*, 153.

42 Botterell and MacLeod, "'Not for the Faint of Heart': Assessing the Status Quo on Parental Licensing," 151.
43 E.J. Graff. "The Lie We Love," *Foreign Policy*, 169 (2008), pp. 58–66.
44 See, for example, Friedrich, "A Duty to Adopt?" 31–32, and Rieder, "Procreation, Adoption, and the Contours of Obligation," 300–301.
45 Friedrich, "A Duty to Adopt?" 32.
46 Friedrich, "A Duty to Adopt?" 32.
47 Liao, *The Right to Be Loved*, 186–187.
48 United Nations Department of Social and Economic Affairs (Population Division), *Child Adoption: Trends and Policies* (New York: United Nations, 2009), 108.
49 Kathryn Joyce, *The Child Catchers: Rescue, Trafficking, and the New Gospel of Adoption* (New York: PublicAffairs, 2013).
50 Graff, "The Lie We Love."
51 Rulli, "Preferring a Genetically-Related Child," 697–698.
52 For an overview with a focus on the Canadian context, see S. Ndegwa and S. Kelly. "Status of Public Funding for In Vitro Fertilization in Canada and Internationally," *Environmental Scan*, 14 (2010), pp. 1–12.
53 De Wispelaere and Weinstock, "Privileging Adoption over ART." For an earlier, similar argument, see Petersen, "The Claim from Adoption."
54 De Wispelaere and Weinstock, "Privileging Adoption over ART," 139.
55 De Wispelaere and Weinstock, "Privileging Adoption over ART," 142.
56 De Wispelaere and Weinstock, "Privileging Adoption over ART," 142. Many other theorists have argued that the current disparity between "natural" and adoptive parents in terms of parental screening is unjustified. See, for example, Botterell and McLeod, "'Not for the Faint of Heart'."
57 Eduardo Rivera-López. "The Claim from Adoption Revisited," *Bioethics*, 20(6) (2006), pp. 319–325, 321.
58 The first three strategies are defended by De Wispelaere and Weinstock in "Privileging Adoption over Sexual Reproduction?" esp. 215–222, whereas the latter is defended by Tina Rulli in "The Unique Value of Adoption."
59 De Wispelaere and Weinstock, "Privileging Adoption over Sexual Reproduction?" 212.

7 Procreation and the Environment

7.1 Introduction

In previous chapters, I have considered how our duties to prospective children and existing children in need of parental care limit the conditions under which it is morally permissible to procreate. In this final chapter, I consider how our duties toward the environment—and to the existing and future people who depend on it—limit the conditions under which it is morally permissible to procreate, as well as the number of children we are morally permitted to create. Faced with the threat of climate change and the risk of harm it presents to present and future people, we often take ourselves to have weighty moral duties to make individual sacrifices and modifications to our lifestyles in order to lower our total level of environmental impact. For example, we take ourselves to have duties to live car-free, to take fewer flights, to adopt a meatless diet, or to improve the energy efficiency of our homes. These are all important measures that individuals might take to lower their total level of impact, though as some commentators have pointed out, the common focus on reducing our consumption ignores what is arguably the most ecologically impactful decision that a person will make in their lifetime: the decision to bring another human being into existence.[1] In choosing to create a child, one is choosing to initiate an additional lifetime of consumption, and perhaps even more if one's child eventually decides to procreate. By some calculations, the impact of this decision in terms of CO_2 emissions is likely to dwarf whatever reductions one makes to their personal level of consumption. For example, in a widely discussed paper calculating the carbon impact of procreation, Paul A. Murtaugh and Michael G. Schlax estimate that the creation of a single child in the United States will add approximately 9,441 tons of CO_2 to the carbon legacy of their parents, or 4,720.5 tons per person. They contrast this figure with the reduction in CO_2 that accompanies other common efforts to reduce our consumption, including increasing the fuel efficiency of our vehicles from 20 to 30 miles per gallon

DOI: 10.4324/9781003164104-8

(148 tons), reducing car travel from 231 to 155 miles per week (147 tons), or replacing single glaze windows with the most energy efficient variety (121 tons). If accurate, these figures suggest that if we really want to reduce our total level of environmental impact, we should strongly consider reducing the number of children we create in addition to reducing our consumption.[2]

Recently, a number of philosophers have drawn on this type of reasoning to support individual moral duties to limit or even refrain from procreation.[3] Many of these philosophers endorse a version of what I will call the *environmental argument for limiting procreation*, which holds that if it is wrong to consume resources or generate environmental impact beyond a certain limit, then, other things being equal, it is also wrong to create however many children it would take to exceed that limit. In this chapter, I explore this argument further with a view to understanding how our climate-related duties impact on the morality of procreation. While I do not disagree that a concern for the environment gives us significant moral reasons to limit the number of children we create—particularly if, as I argued in Chapter 1, the core interest in parenting can be satiated with the creation of just one child—I argue that proponents of the environmental argument for limiting procreation have failed to demonstrate that we are under a moral duty to do so, or that we necessarily act wrongly by creating children in excess of some particular number. I focus my arguments on what I take to be three key limitations of this argument. First, I argue that many versions rely on implausible and unworkable conceptions of how much of a child's consumption is legitimately attributable to their procreative parents, leading them to overestimate the ecological impact of procreation. Second, I argue that all versions have failed to defend specific and independently justified accounts of the limits of permissible consumption and hence have not demonstrated that the creation of some particular number of children would cause a procreative parent to exceed that limit. Finally, I argue that all versions fail to acknowledge important qualitative differences between procreation and other forms of consumption that might justify treating them differently from a moral point of view.

The chapter proceeds in four parts. In Section 7.2, I outline the basic structure of the environmental argument for limiting procreation and explain its dependence on a number of key variables, including how we understand the limits of permissible consumption and how we count the consumption necessitated by procreation against that limit. In Section 7.3, I consider the latter variable; in Section 7.4, I consider the former. Finally, in Section 7.5, I consider arguments for rejecting a key assumption of the environmental argument for limiting procreation, namely, the notion that procreation and consumption ought to be treated equivalently from a moral point of view.

7.2 The Environmental Argument for Limiting Procreation

In an influential paper, Thomas Young argues that "it is inconsistent to believe that overconsumption is wrong or bad yet believe that having children is morally permissible, insofar as they produce comparable environmental impacts, are voluntary choices, and arise from similar desires."[4] To illustrate the plausibility of this claim, Young compares the environmental impact produced by two hypothetical couples living in an affluent society: the *Greens*, a childless couple in their early thirties who are expected to live into their early eighties; and the *Grays*, who are identical to the Greens in terms of their age and life expectancy, but who plan to create two additional children during their lifetime. According to Young, if we assume a fixed per-person level of consumption, the Grays can be expected to produce over 2.5 times the environmental impact as the Greens. His reasoning for this runs as follows. Both of the Greens are expected to live for 50 more years and will therefore collectively produce 100 years' worth of environmental impact. The Grays will produce the same level of impact during those years, but they will also be responsible for the additional impact produced by their two children. Supposing that both of their children will also live for 80 years, the Grays will produce 260 years of environmental impact, or just over 2.5 times that produced by the Greens.

According to Young, the case of the Greens and the Grays presents a dilemma for "mainstream environmentalists" who want to maintain both that procreation is morally permissible and that overconsumption is morally wrong. Few environmentalists would likely criticize the Grays for pursuing their seemingly benign procreative projects. However, once we take into account the environmental impact of procreation, this attitude becomes difficult to square with the view that consuming excessive amounts of resources is morally wrong. Imagine, for example, that the Greens were to dramatically increase their consumption to a level equivalent to the Grays by adopting a lifestyle of "eco-gluttony," wherein they engage in a level of consumption, resource depletion, and waste production that far exceeds that of the average person in their society. Many environmentalists would condemn the Greens for adopting a lifestyle with such a high environmental impact, but if this is the attitude taken toward the Greens, then consistency seems to demand that the same attitude be extended to the Grays. This is supported further by the fact that each couple's respective lifestyle decision also shares other important features, including the fact that they are both undertaken voluntarily and arise from similar desires (e.g., to achieve happiness or pursue their own conceptions of the good life). In light of these similarities, Young concludes that decisions about consumption and procreation ought to be treated equivalently from a moral point of view. Following others in the literature, I will refer to this view as the *moral equivalence thesis*.[5]

It is important to note that, taken on its own, the moral equivalence thesis does not imply anything about the moral permissibility of either procreation or excessive consumption—it only implies that we ought to treat them equivalently from a moral point of view. It is perfectly consistent, then, to endorse the moral equivalence thesis while believing that both procreation and excessive consumption are morally permissible. However, when paired with the assumption that it is wrong to individually consume resources beyond a certain limit, the moral equivalence thesis implies that it is also wrong to create however many children it would take to exceed that limit. We can express this implication in terms of the following premises, which I will refer to as the *environmental argument for limiting procreation*:

1 It is morally wrong to individually consume resources beyond a certain limit due to its adverse environmental impact;
2 Being responsible for the consumption of resources as a result of the voluntary decision to procreate is morally equivalent to individually consuming resources; therefore,
3 It is morally wrong to create however many children it would take to exceed the limit of permissible consumption indicated in (1).

Several philosophers—who, following Kalle Grill, I will collectively refer to as *procreative limitarians*[6]—have drawn on this type of argument to support individual duties to limit or even refrain from procreation, at least among those living in affluent societies with high per capita levels of consumption. These include duties to refrain from procreation entirely,[7] to have no more than one child per couple,[8] to have no more than one child per person (or two children per couple as in replacement-level procreation),[9] or to limit oneself to some unspecified number of children.[10] Of course, the particular limits on procreation this argument is able to support will depend on how we understand the limits of permissible consumption indicated in premise (1), as well as how we count the consumption necessitated by procreation against that limit, both of which may vary considerably based on a range of different factors, including how we understand the wrongness of individual overconsumption and what types of consumption we consider people to be morally responsible for. I will examine these issues in Sections 7.3 and 7.4 as a means of better understanding what the environmental argument for limiting procreation might actually entail, on the assumption that the moral equivalence thesis expressed in premise (2) is true. I will work backward here, first determining the extent to which the consumption necessitated by procreation should be understood as part of a procreator's total level of consumption and then examining what type of limits on consumption procreative limitarians must endorse in order to support ecological limits on procreation. Then, in Section 7.5, I will consider arguments for rejecting the moral equivalence thesis on the grounds that procreation and consumption are relevantly distinct and should not be treated equivalently from a moral point of view.

7.3 Procreation and Responsibility for Environmental Impact

The judgment that the Greens act wrongly by increasing their level of consumption presupposes there are limits on the total amount of resources that individuals are morally permitted to consume or the total amount of environmental impact they are morally permitted to create. If the Greens act wrongly by adopting a lifestyle of eco-gluttony, it is because their level of consumption exceeds this limit, thereby rendering it morally objectionable *overconsumption*. It is of course an open question whether the Greens actually exceed this limit by increasing their level of consumption, and hence whether the Grays exceed this limit by creating two children, as this will ultimately depend on what we determine this limit to be (about which I will have more to say in Section 7.4). What I would like to consider first, however, is how the environmental argument for limiting procreation conceptualizes the impact of procreation in relation to this limit. If procreation is morally equivalent to consumption in virtue of its environmental impact, then, just as the impact of one's decisions about consumption must count toward the limits of permissible consumption, so too must the impact of one's decisions about procreation. Another way of putting it is to say that, on the terms of the environmental argument for limiting procreation, one's total level of environmental impact not only includes one's personal consumption but also at least some of the consumption of one's children that is necessitated by the decision to procreate.

These ideas are implicit in Young's argument, though they are presented more explicitly by other procreative limitarians. For example, arguing along similar lines as Young, Corey MacIver insists that we must "collapse the dichotomy between consumption and procreation, and acknowledge procreation as one of the most significant ways in which individual human beings lay claim to the natural world."[11] According to MacIver,

> When we assess the ecological impact of a given individual (or, perhaps more appropriately, a given set of biological parents), this assessment should include not only their demands upon the natural world, but also their contribution to the creation of new demands on the natural world through procreative means.[12]

MacIver goes on to explicitly link this assessment of a person's ecological impact to their resource entitlements, claiming "Like other claims to natural goods, the claims associated with procreation should be understood both as part of the ecological impact for which the procreator is responsible and as subtractable from any entitlement the procreator may have."[13] Let's call this type of view the *responsibility thesis*, which we can express generally as follows:

> *Responsibility Thesis*: the environmental impact for which a person is morally responsible includes their own consumption in addition to at least some of the consumption of any children they are morally responsible for creating.

Procreative limitarians are committed to the responsibility thesis in one form or another. But is it defensible? This will depend critically on *how much* of children's consumption we consider their procreative parents to be morally responsible for, of which there are different possible views. Let's begin, then, by considering the strongest of these views, which I will call the *total lifetime consumption* view:

> *Total Lifetime Consumption*: the environmental impact for which a person is morally responsible includes (a) the totality of their own consumption, and (b) the totality of the consumption of any child they are morally responsible for creating, divided by their level of responsibility for that child's existence.

The total lifetime consumption view is presupposed in Young's analogy of the Greens and the Grays, wherein the Grays are deemed to be jointly responsible for the totality of their children's consumption (recall that the 80 years of environmental impact represented by each of the Gray children are added on to the Grays' total level of environmental impact). It is also explicitly endorsed by Murtaugh and Schlax in their calculation of the carbon impact of procreation.[14] The basic assumption informing their methodology is that

> a person is responsible for the carbon emissions of his descendants, weighted by their relatedness to him … So, for example, a mother and father are each responsible for one half of the emissions of their offspring, and ¼ the emissions of their grandchildren.[15]

When viewed in a purely forward-looking manner, the total lifetime consumption view might seem plausible. After all, the consumption of the Gray children would not have occurred but for the Grays' voluntary decision to procreate, so it seems reasonable to count this consumption as part of the impact for which the Grays are responsible (and hence as subtractable from *their* permissible consumption limit). However, while this type of view might seem reasonable when considered from the perspective of a single generation, it begins to face complications when applied intergenerationally. If parents are jointly responsible for the totality of their children's consumption, then this carries one of two implications for the present generation of parents: either (a) the present generation of parents are not responsible for their own consumption, including the consumption necessitated by their decisions to procreate, as this is considered part of the environmental impact for which *their* parents are morally responsible (which in turn would be considered part of the environmental impact for which *their* parents are morally responsible and so on until the first generation); or (b) the present generation of parents are in fact responsible for their own consumption as well as a proportion of the consumption of *everyone* who comes into existence as a result of their decision to procreate, including their children, grandchildren, great grandchildren, and so on *ad infinitum*.

Both of these implications are deeply problematic for procreative limitarians. Implication (a) undermines the very principle of personal responsibility on which the environmental argument for limiting procreation is based. Rather than holding individuals responsible for their own consumption choices, including their procreative choices, (a) appears to support a form of intergenerational buck passing, wherein responsibility for consumption is always attributable to a previous generation. If the total lifetime consumption view has this implication, it is not clear that it can support *any* limits on procreation based on a duty to limit one's consumption, as it does not actually hold people responsible for their own consumption past an initial generation of procreators. Implication (b) avoids this problem by holding persons responsible for their own consumption in addition to a proportion of the consumption of everyone who comes into existence as a result of their decision to procreate, though in so doing, it appears to place severe, potentially anti-natal restrictions on the moral permissibility of procreation, insofar as the limits of permissible consumption cannot accommodate one's own consumption in addition to a proportion of the consumption of a potentially limitless line of descendants. This, however, is absurd: it is primarily out of respect for the interests of future persons that we have duties to limit our consumption in the first place, so discharging these duties cannot require a course of action that would prevent those persons from ever existing (lest the duties themselves disappear).

The total lifetime consumption view is therefore untenable. MacIver appears to recognize this problem, noting that considerations of personal responsibility seem to weigh against counting *all* of children's consumption as part of the impact for which their parents are morally responsible. As he claims,

> None of the impacts associated with my existence would occur had my parents not decided to bring me into existence, but it is important to acknowledge that I bear responsibility for making better and worse choices (including my own procreative choices) as I go through life. Any allocation of ecological footprint, and hence of moral responsibility for environmental impact, should reflect the agency of both parents and children (or rather, of the adults that children will become).[16]

According to MacIver, an appropriate focus on agency and moral responsibility will lead us to one of two views about how much of a child's consumption is legitimately attributable to their parents. On the first, weaker view, parents are responsible for the level of consumption that is required for their children to live minimally decent lives, as this level of consumption is not optional for their children. However, they are not responsible for the non-essential consumption their children will engage in once they develop the capacity for agency and moral responsibility, including their own decisions to procreate. Let's call this version of the responsibility thesis the *essential consumption* view:

> *Essential Consumption*: the environmental impact for which a person is morally responsible includes (a) their own non-essential consumption, (b) the essential consumption of any child they are morally responsible for creating, and (c) that child's non-essential consumption prior to developing the capacity for agency and moral responsibility, divided by their level of responsibility for that child's existence.[17]

On the second, stronger view, parents are responsible not only for the level of consumption that is required for their children to lead minimally decent lives but also for the non-essential consumption that is *reasonably foreseeable* at the time of procreation. In contemporary North America, for example, it is reasonably foreseeable that one's child will engage in a significant level of non-essential consumption during their lifetime, including the consumption of vast amounts of energy, as the social entrenchment of that level of consumption makes it extremely difficult and costly to avoid (e.g., by "going off the grid"). Thus, MacIver concludes that parents may be responsible "not only for the emissions that are literally unavoidable, but also for the foreseeable emissions that contingent circumstances make very difficult or excessively demanding to avoid."[18] Let's call this version of the responsibility thesis the *reasonably foreseeable consumption* view:

> *Reasonably Foreseeable Consumption*: the environmental impact for which a person is morally responsible includes (a) their own consumption that was not foreseeable by their parents, (b) the essential

consumption of any child they are morally responsible for creating, and (c) that child's non-essential consumption that is reasonably foreseeable both before and after they develop the capacity for agency and moral responsibility, divided by their level of responsibility for that child's existence.

Do either of these views provide a more plausible alternative to the total lifetime consumption view? Let's consider them in reverse order. If parents are responsible for the level of consumption that is reasonably foreseeable at the time of procreation, then it is difficult to see how the reasonably foreseeable consumption view can avoid the challenges associated with the total lifetime consumption view, for the simple reason that it is reasonably foreseeable that one's children may eventually decide to procreate. This leads to a position that is almost identical to the total lifetime consumption view: in addition to being responsible for the consumption their children require to live minimally decent lives, parents are also responsible for their children's reasonably foreseeable non-essential consumption past the point at which they develop the capacity for agency and moral responsibility, including the consumption necessitated by their own decisions to procreate. Either parents are responsible for a significant proportion of this total level of consumption, in which case the creation of *any* children may cause them to exceed the limits of permissible consumption, or responsibility is attributable to a previous generation, in which case they face virtually no limits on either consumption or procreation.

The essential consumption view has the potential to avoid these problems by holding children responsible for their own non-essential consumption once they develop the capacity for agency and moral responsibility. It therefore provides a cut-off point at which parents are no longer responsible for a significant portion of their children's total consumption, including the consumption necessitated by their own decisions to procreate.[19] If we apply this view to the case of the Greens and the Grays, we arrive at a much different calculation than Young about the total amount of impact that each couple is responsible for. Recall that, on Young's calculation, the Greens will be responsible for 100 years of environmental impact (50 years each), whereas the Grays will be responsible for 260 years of environmental impact (50 years each, plus 80 years for each of their children). However, on the essential consumption view, we need to subtract from each couple's total level of impact the consumption for which they cannot be held morally responsible. For the purposes of illustration, let us assume that one acquires the capacity for full agency and moral responsibility at around 18 years of age and that the level of consumption required for a minimally decent life accounts for around 25% of the average person's consumption in the particular society in which the Greens and the Grays live.[20] We would then need to subtract 25 years of impact from both the Greens and the Grays, as they cannot be held responsible for consuming the resources that are required for them to live minimally decent lives for their remaining 50 years (totaling 12.5 years each), and we would also need to subtract from the Grays' total impact the portion of their children's impact for which they cannot be held morally responsible. Because each Gray child is expected to live for 80 years, each will be morally responsible for 62 years of impact (80 – 18 years) minus the unavoidable consumption that is necessary for them to live minimally decent lives during that time (62 × 0.25 = 15.5 years), for a total of 46.5 years of impact each. The Grays will be responsible for the remainder (33.5 × 2 = 67 years) in addition to their own non-essential consumption during that time (75 years), for a total of 142 years of impact relative to the Greens' 75 years. Notice that this total is significantly lower than Young's total in both absolute and comparative terms, representing just over half of the total impact that Young attributes to the Grays and less than two times the total impact of the Greens (as compared to more than two and a half times the impact on Young's formulation).

The contrast between these views becomes even more apparent when we replace Young's purely comparative metric of "years of consumption" with the more concrete and commonly used metric of CO_2 emissions. I noted earlier that Murtaugh and Schlax have provided the most widely discussed account of the carbon impact of procreation, wherein they explicitly endorse the total lifetime consumption view. On their calculation, if we assume (a) that a person is responsible for the carbon emissions of his descendants, weighted by their relatedness to him, and (b) that emissions and fertility rates will remain relatively constant, then creating one child in the United States will add approximately 9,441 tons of CO_2 to the carbon legacy of their parents, or 4720.5 tons of CO_2 per person.[21] However, if we endorse the essential consumption view and assume that a procreator is not responsible for their children's non-essential consumption after they develop the capacity for agency and moral responsibility, including the consumption necessitated by their own decisions to procreate, then this figure drops dramatically. For example, Felix Pinkert and Martin Sticker estimate that individual subsistence emissions can be calculated at 1.4–4.4 tons of CO_2 per year in developed countries, suggesting that a person who lives for 80 years will produce between 112 and 352 tons of CO_2 in subsistence emissions. If we add to this total the non-essential emissions for which they are not morally responsible during the first 18 years of their life, we get a total somewhere in the area of 328–568 tons of CO_2.[22] When we divide this value equally between two procreators, this means that, on the terms of the essential consumption view, creating a child will add between 164 and 284 tons of CO_2 to an individual's carbon footprint, as opposed to 4720.5 tons of CO_2 on the total lifetime consumption view. If this is correct, it suggests that forgoing the creation of one child is actually on par with other significant efforts that individuals might take to reduce their total emissions, including living car-free or adopting a meatless diet, which can reduce one's emissions by 192 and 72 tons of CO_2, respectively, over the course of an 80-year lifespan.[23]

This conclusion might seem surprising and even obviously mistaken—after all, by refraining from having a child, one is preventing the total emissions of that child in addition to any number of possible descendants, whereas by refraining from driving a car, one is only preventing the emissions associated with one's own car usage. However, it is important to remember that what we are accounting for here are not the total emissions for which one is causally responsible, but rather the total emissions for which one can legitimately be held morally responsible. Procreators cannot be held morally responsible for *all* of the emissions of their children, as a principle that required this would be inconsistent with holding them responsible for their *own* emissions, including those resulting from their decisions to procreate, as these emissions would then be attributable to *their* parents (which would, in turn, be attributable to their parents' parents and so on until the first generation). Thus, procreative limitarians must endorse a version of the responsibility thesis that somehow divides responsibility for emissions between procreators and their children in a way that avoids this type of regress. The essential consumption view achieves this by only holding procreators responsible for their children's essential emissions once they develop the capacity for agency and moral responsibility and thus providing a cut-off point at which they are no longer responsible for a significant portion of their children's total emissions, including the emissions associated with their own decisions to procreate. In so doing, however, it also attributes a much smaller impact value to procreation than procreative limitarians have tended to assume. While this version of the responsibility thesis sits uneasily with claims of some procreative limitarians, who want to maintain that "procreative decisions are *by far* the most ecologically significant decisions that individuals will make in their lives,"[24] it is seemingly the only version available to them given the significant problems associated with the other two.

7.4 How Many Children Is Too Many?

So far I have argued that proponents of the environmental argument for limiting procreation are committed to a version of the responsibility thesis and that the only tenable version is the essential consumption view. This view holds procreators responsible for the essential consumption of any children they create, divided by their level of responsibility for that child's existence, but does not hold them responsible for their children's non-essential consumption after they develop the capacity for agency and moral responsibility, including the consumption necessitated by their children's own decisions to procreate. By exempting procreators from responsibility for the procreative decisions of their children, the essential consumption view is able to avoid the problems associated with the total lifetime consumption and reasonably foreseeable consumption views, but in so doing assigns a much smaller impact value to procreation than these alternatives (which many procreative limitarians happen to endorse, either explicitly or implicitly).

Of course, "smaller" is a relative term, and even though the essential consumption view assigns a much smaller impact value to procreation than the total lifetime consumption or reasonably foreseeable consumption views, it is important to acknowledge that creating a child will still rank among the most ecologically impactful decisions that a person can make in absolute terms, particularly for those living in affluent societies with high per capita levels of consumption. Does this entail that individuals are morally obligated to limit the number of children they create? And if so, how many children (if any) must they limit themselves to?

Our answers to these questions will depend on how we conceptualize what I have been calling the limits of permissible consumption, which indicate how much an individual is morally permitted to consume or how much environmental impact they are morally permitted to create. On the terms of the environmental argument for limiting procreation, it is morally wrong to create a child if and only if doing so would cause one to exceed the limits of permissible consumption. Thus, when a proponent of the environmental argument for limiting procreation concludes that it is morally wrong to create *n* children, they are committing to the following claim:

$$\mathrm{P}n + \mathrm{C} > \mathrm{L}$$

P = Environmental Impact of Procreation
n = Number of Children
C = Individual Consumption for Which One Is Morally Responsible
L = Limits of Permissible Consumption

In order to assess the plausibility of this claim, we need to fill in the variables. If the arguments in the previous section are correct, then the value of P should be determined by adding the essential consumption of a child over their expected lifespan to their non-essential consumption prior to developing the capacity for agency and moral responsibility, divided by the procreator's level of responsibility for that child's existence, and the value of C should be determined by adding up the procreator's non-essential consumption after the point at which they develop the capacity for agency and moral responsibility. There is bound to be controversy regarding what these values actually are due to reasonable disagreement about what counts as essential consumption or when a person develops the capacity for agency and moral responsibility, but we at least know *how* to go about determining them for the purpose of assessing whether $Pn + C > L$.

But how should we go about determining L? This is significantly more complex. One reason for this complexity is general skepticism about individual duties to limit consumption. For example, Walter Sinnott-Armstrong denies that there are individual duties to reduce CO_2 emissions on the grounds that our individual emissions are inconsequential to the harms associated with climate change.[25] To support this view, he imagines how we might object to a token instance of frivolous consumption, such as joyriding in a gas-guzzling SUV. It is natural to think that such an act is wrong and that it is wrong because of its contribution to the harms of climate change, though in Sinnott-Armstrong's view, this diagnosis rests on an implausible conception of what it means to count as the cause of a harm. Arguably, in order to count as the cause of a harm, an action or decision must be a necessary and/or sufficient condition of that harm, though joyriding in an SUV satisfies neither condition: the harms of climate change are just as likely to occur if one does not joyride in the SUV (so it is not necessary) and joyriding in the SUV will not by itself bring about the harms of climate change (so it is not sufficient). Sinnott-Armstrong's point is not that there are *no* obligations to reduce greenhouse gas emissions, but rather that these obligations attach primarily to more causally efficacious actors, such as states. On this view, individuals may have duties to support governments and policies that are committed to a reduction in greenhouse gas emissions, but they do not have moral duties to reduce their individual emissions.[26] Applied in the present context, this argument implies that individuals may have duties to support justice-respecting laws or social policies that will result in a reduction in the birth rate and overall population (e.g., through the advancement of gender equality and the universal provision of contraception and sex education),[27] but they do not have individual duties to limit the number of children they create.

How might a procreative limitarian respond to this challenge? One possible response is to reject the assumption that our individual consumption decisions are causally inconsequential. Sinnott-Armstrong assumes that, with respect to the harms of climate change, our individual consumption decisions are but a drop in the bucket (or, to bring the analogy to scale, but a drop in a river that is going to flood downstream due to torrential rains).[28] In one sense, this seems difficult to deny—as Chad Vance has recently pointed out, the average American will produce about 1,135 tons of CO_2 over the course of their entire lifetime, whereas about 1,151 tons of CO_2 are emitted globally every second, suggesting that it is simply "not plausible that our individual contributions to climate change make any perceptible difference to the harm that occurs."[29] However, some philosophers have pushed against this view by offering alternative ways of conceptualizing the harm associated with individual emissions. For example, John Nolt concedes that each American alive today will account for only a tiny proportion of global emissions during their lifetime, but because the effects of these emissions will be experienced for many years into the future by billions of people, this tiny proportion may turn out to represent serious harm. Assuming that the average American is responsible for one billionth of current and near-term emissions and that these emissions will result in serious harm to at least 4 billion people over the next millennium, Nolt estimates that "the average American causes through his/her greenhouse gas emissions the serious suffering and/or deaths of two future people."[30] If these estimates are correct, they offer an explanation of how premise (1) of the environmental argument for limiting procreation can be true despite the fact that our individual consumption choices are merely a drop in the bucket.

Of course, this type of response depends on these estimates being correct, though as Nolt himself concedes, they are ultimately crude calculations that "could easily be wrong by an order of magnitude, maybe more, in either direction."[31] There may also be additional challenges with appealing to the notion of harm in order to explain the wrongness of individual overconsumption. For example, some worry that such an approach cannot distinguish between subsistence and luxury consumption given that both contribute equally to the harms of climate change, and so cannot explain the distinctive wrongness of consuming resources beyond a particular limit.[32] Thus, a second and more promising line of response to Sinnott-Armstrong's challenge is to identify a principle that can explain the wrongness of individual overconsumption without assuming that it directly results in serious harm to others. One such principle is based on considerations of distributive justice. If human beings must collectively limit their level of consumption or environmental impact in order to ensure a sustainable world for future generations, then the permission to consume resources or generate environmental impact is a scarce resource that ought to be distributed fairly between individuals.[33] This implies that there is some level of consumption that each individual is morally permitted to engage in while fulfilling their duties to future generations. When an individual exceeds this limit by engaging in excessive consumption, they are acting wrongly in at least two ways, even if their actions do not directly or immediately result in serious harm to others: first, they are failing to perform their share of a collective duty owed to members of future generations, and in so doing, disregarding their interests and risking a state of affairs in which that duty is not fulfilled[34]; and second, they are failing to respect the just claims of their contemporaries, for by consuming more than their fair share of resources, they are diminishing the share that is available to others and thus putting them into a position where claiming their own fair share may become incompatible with fulfilling their own portion of the collective duty owed to members of future generations.[35]

I will assume in what follows that this type of account provides a reasonable explanation of why it is morally wrong to individually consume resources beyond a certain limit, and hence that general skepticism about the limits of permissible consumption is unfounded (indeed, it seems impossible to make sense of the notion of individual *over*consumption unless we assume that such limits exist). However, in order to determine whether P*n* + C > L, we need a specific and independently justified account of what these limits actually are. This is where a second layer of complexity reveals itself, as advancing an independently justified account of L requires engaging with a range of normative and empirical issues about which there is both deep disagreement and a considerable degree of uncertainty. Simon Caney gives a sense of this complexity in the context of a view he calls *ecological liberalism*.[36] Like procreative limitarians, Caney endorses the moral equivalence of procreation and consumption and believes that, like the impact of one's decisions about consumption, the impact of one's decisions about procreation ought to be subtracted from one's ecological resource entitlements. However, he is skeptical of the idea that this supports any particular limit on the number of children one might create, as this will depend on an individual's total level of consumption: individuals who consume less (such as those in the Global South) will be able to "afford" to have more children while living within their ecological limits, while those who consume more (such as those in the Global North) will be able to "afford" to have fewer children while living within those limits. What is important for Caney is not that individuals limit themselves to the creation of some specific number of children, as philosophers like Sarah Conly and Christine Overall have argued, but rather that they live within certain ecological limits and bear responsibility for their choices.[37] As long as they do so, Caney insists that they should have the freedom to choose how they go about complying with those limits.

For present purposes, however, the important aspect of Caney's view is the methodology he proposes for determining what these limits are. According to Caney, determining the limits of permissible consumption necessarily involves consideration of multiple variables. First, we must determine what our moral responsibilities are to members of future generations, as these will inform the upper limit of the resources the present generation is morally permitted to consume or the total level of environmental impact they are morally permitted to create. Philosophers disagree about the nature and extent of these responsibilities. For example, some endorse an egalitarian conception of these responsibilities, according to which we are obligated to secure the environmental preconditions for members of future generations to enjoy a standard of living that is no worse than the standard we currently enjoy; others endorse a sufficientarian conception, according to which we are obligated to secure the environmental preconditions for members of future generations to live minimally decent lives; and others endorse different conceptions altogether. Each of these views will impose a set of duties on the present generation that is more or less demanding, so before we can assess what discharging our responsibilities to future generations requires of the present generation, we first need to arrive at a conception of what future generations are entitled to as a matter of justice.

Once we settle on a conception of intergenerational justice, the second step is to draw on the empirical literature concerning the environmental impact of human activity and the environmental preconditions of realizing justice in order to determine the ecological limits within which the present generation must live. For example, in their 2023 Synthesis Report, the Intergovernmental Panel on Climate Change (IPCC) noted that in order to keep the global average temperature rise beneath 2°C above pre-industrial levels, and thus avoid the most catastrophic effects of climate change, the world must limit itself to the emission of 1,500 gigatons of CO_2 ($GtCO_2$); in order to keep the global average temperature rise beneath 1.5°C above preindustrial levels and mitigate the impacts of climate change even further, the global carbon budget drops to 500 $GtCO_2$.[38] Depending on our conception of intergenerational justice, these figures may indicate the ecological limits within which the present generation must live in order to fulfill their responsibilities to future generations.

Finally, once we determine the ecological limits within which the present generation must live, the third step is to determine how the burdens associated with adhering to those limits ought to be distributed among members of the present generation. One might think that we can arrive at a fair distribution simply by dividing the global carbon budget by the world's population, as in the so-called *equal per capita* view. Suppose, for example, that we adopt a weaker account of our obligations to future generations and determine that we must limit ourselves to the emission of 1,500 $GtCO_2$ so as to avoid only the most catastrophic effects of climate change. With a global population of approximately 8.025 billion people as of 2023, this would result in an individual carbon budget of around 186.9 tons of CO_2 per person. However, an equal distribution is not necessarily a fair distribution, and there are good reasons to think that fairness might require departing from the equal per capita view. One such reason is the fact that individuals are situated differently with respect to the level of carbon emissions that are necessary for them to achieve a decent standard of living, while another is the fact that individuals are situated differently with respect to how they have benefitted from past emissions. Accounting for these differences may require a drawing on a different distributive principle, which will in turn alter the size of individual consumption limits.

These are just some of the variables influencing how we conceptualize the limits of permissible consumption, though taken together, they demonstrate that arriving at a specific and independently justified account of what these limits are is no simple task. At the very least, this shows that existing versions of the environmental argument for limiting procreation are in one sense incomplete, for while each has drawn attention to the significant ecological impact associated with procreation, none have defended accounts of the limits of permissible consumption and have not therefore demonstrated that P*n* + C > L. As in Young's argument, these limits have normally been assumed but not argued for in the context of comparative claims about the ecological impact of procreation relative to other high-consumption activities, where proponents conclude that if it is wrong to engage in a particular class of high-consumption activities, then, other things being equal, it must also be wrong to create the number of children it would take to match that level of consumption. This is not to say that a case cannot be made for the claim that P*n* + C > L, but rather that such a case has generally not been made and that making one is bound to be fraught with controversy given the difficulty involved in defending an account of the limits of permissible consumption.

To complicate matters further, however, there is an important variable that Caney overlooks in his proposed methodology for determining the limits of permissible consumption, which adds a third layer of complexity to our assessment of whether P*n* + C > L. In addition to our conception of intergenerational justice, the extent of our responsibilities to future generations will also depend crucially on the *size* of the population that will exist in the future, which is of course a function of the procreative decisions we make today. Suppose, for example, that we endorse an egalitarian conception of intergenerational justice and assume that the present generation is obligated to secure the environmental preconditions for members of future generations to enjoy a standard of living that is no worse than the standard we currently enjoy. What discharging this obligation requires in the way of limiting consumption or environmental impact will vary considerably depending on how many people will exist in the future—for example, securing equal conditions for future generations in a scenario where the population size doubles will be vastly more demanding than a scenario in which the population size remains constant—yet how many people will exist in the future will be determined in part by the procreative decisions made by the current generation, which our duties of intergenerational justice are meant to constrain. This introduces an aspect of circularity into the environmental argument for limiting procreation: in order to determine how many children we are morally permitted to create, we need a specific and independently justified account of the limits of permissible consumption; however, in order to arrive at a specific and independently justified account of the limits of permissible consumption, we need an idea of how many people will exist in the future. It seems, then, that in addition to significant complexity regarding how to determine the limits of permissible consumption, the environmental argument for limiting procreation also faces a chicken-and-egg problem.

Perhaps it is possible to circumvent this problem in a way that would allow the argument to work. For example, if we are positing an egalitarian conception of intergenerational justice, proponents of the environmental argument for limiting procreation might identify a maximum future population size for which we are able to secure an equal standard of living and then calculate the limits of permissible consumption on that basis. Notice, however, that in taking this approach, proponents of the environmental argument for limiting procreation would betray a central commitment of that argument, namely, that procreation and consumption ought to be treated equivalently from a moral point of view. Indeed, if the limits of permissible consumption are determined on the assumption of the maximum population size, then each individual's resource entitlements will have been calculated so as to accommodate the creation of the maximum possible number of children. But this is exactly the kind of privileging of procreation over other forms of consumption that procreative limitarians reject.

Where, then, does this leave proponents of the environmental argument for limiting procreation? It seems like they face at least three options. The first option is to navigate these complexities and proceed with defending the view that $Pn + C > L$. To do so, they will have to defend an independently justified account of L that takes into consideration (a) our duties toward future generations, (b) the ecological limits they impose on the present generation, (c) the just distribution of the burdens associated with adhering to those limits among individuals, and (d) the size of the population that will exist in the future. They will then have to demonstrate that this value is in fact lower than the impact of the creation of n children when calculated on the basis of the essential consumption view.

The second option is to replace premise (1) of the environmental argument for limiting procreation with a different type of principle that does not require proponents to demonstrate that $Pn + C > L$. As I have presented it above, premise (1) of the environmental argument for limiting procreation holds that it is morally wrong to individually consume resources beyond a certain limit due to its adverse environmental impact. This principle invites the question of what this limit is, which in turn requires defending positions on variables (a)–(d) listed above. However, the difficulty associated with determining the limits of permissible consumption might provide grounds for endorsing a different type of principle. In cases where there is deep uncertainty about the impacts and risks associated with a particular course of action, it is often reasonable to err on the side of caution and endorse something like the precautionary principle. Endorsing such a principle in the present context may require individuals to limit their consumption to the greatest extent possible while maintaining a minimally decent standard of living. This would yield the following revised version of the environmental argument for limiting procreation:

1 We are morally obligated to limit our individual consumption to the greatest extent possible while maintaining a minimally decent standard of living;
2 Being responsible for the consumption of resources as a result of the voluntary decision to procreate is morally equivalent to individually consuming resources; therefore,
3 We are morally obligated to limit ourselves to the fewest number of children that is consistent with maintaining a minimally decent standard of living.

This revised version of the argument presents at least two advantages for procreative limitarians. First, it allows them to bypass the difficult issue of providing a general account of the limits of permissible consumption, which requires defending positions on variables (a)–(d) listed above. The upper limit of permissible consumption for any particular person will simply be the minimum level of consumption required for them to maintain a minimally decent standard of living, which may vary by individual depending on their personal and social circumstances. There will certainly be controversy surrounding what constitutes a minimally decent standard of living, but defending an account of this is much more manageable than defending a general account of the limits of permissible consumption.

Second, this revised version of the argument is also better equipped to support non-arbitrary limits on the number of children individuals are morally permitted to create and can therefore supply an answer to the question, "how many children is too many?" There are in fact two distinct answers this revised argument can support, depending on the position we take on whether the opportunity to create a child is necessary for a minimally decent life. If we take the position that the opportunity to create a child is *not* necessary for a minimally decent life, then the fewest number of children required to live a minimally decent life is zero, in which case the revised environmental argument for limiting procreation will not permit the creation of *any* children. Some procreative limitarians appear to endorse this extreme position, though there are several reasons to reject it. First, as I pointed out earlier in connection with the total lifetime consumption view, the duty to limit our consumption is premised on the assumption that people will exist in the future who can be negatively impacted by our present consumption patterns, so discharging this duty cannot require a course of action that would prevent those people from ever existing. If everyone refrained from procreating, there would be no future people; however, if there were no future people, there would be no basis for the duty to limit our consumption (or at least much less of a basis, as we would then only have to consider the impact of our consumption patterns on the interests of presently existing people). Second, even if one could maintain a decent standard of living without personally creating a child, it is doubtful that one could do so in a scenario in which *everyone* refrained from creating a child. This is due to the importance of what Samuel Scheffler has termed the "afterlife conjecture," or the assumption that humanity will continue to live on after we die.[39] If there was no collective afterlife, Scheffler believes that we would cease to value many of what we take to be our most important aims, including medical or scientific research, social or political activism, environmental protection, or the development of arts and culture. It is difficult to imagine that life could be minimally decent in this type of scenario, and hence there is good reason to believe that the revised version of the environmental argument for limiting procreation must permit the creation of at least some minimum number of children (which may be less than one per person) even if we assume that creating a child is not necessary for a minimally decent life.

If, on the other hand, we take the position that the opportunity to create a child *is* necessary for a minimally decent life, then the fewest number of children consistent with maintaining a minimally decent standard of living is seemingly one. Many commentators regard this as a plausible position given the distinctive value associated with the creation of one child versus the creation of multiple children. For instance, Conly argues that whatever value is associated with procreation can be achieved by procreating just once,[40] whereas Tim Meijers has argued that "the right to have one child and the right to have multiple children are fundamentally different. The first child gives access to parenthood, whereas the second child does not: the procreators are already parents."[41] Of course, if, as I argued in Chapter 1, procreation is valuable primarily as a means of initiating a parent-child relationship, and the value of parent-child relationships is conceived independently of any biological connection between parent and child, then we might think the revised version of the environmental argument of limiting procreation actually supports a procreative limit of *less* than one child, to the extent that one can initiate such a relationship with an existing child in need of parental care.

Of course, while this revised version of the argument presents the advantage of supporting specific and non-arbitrary limits on the number of children we are morally permitted to create, it also comes with a significant challenge, as the precautionary principle defended in premise (1) is extremely demanding. Limiting our individual consumption to the greatest extent possible while maintaining a minimally decent standard of living will not only preclude us from creating multiple children but also preclude us from engaging in a wide range of activities that we normally consider to be morally innocuous, from taking a family vacation overseas, to driving our children to soccer practice, to taking a long, relaxing shower at the end of our workday. In this sense, the cost of supporting specific and non-arbitrary limits on the number of children we are morally permitted to create seems to be the adoption of an extremely demanding and by some accounts implausible account of our individual climate-related duties.[42]

This suggests a third option for procreative limitarians, which is to dispense with the language of duties altogether and reframe their arguments in terms of other, less rigid moral considerations. Travis Rieder is an example of a philosopher who has taken this approach.[43] According to Rieder, the various difficulties associated with appealing to individual moral duties in the context of climate ethics should lead us to embrace other moral considerations, such as character virtues and reasons for action. For example, even if we do not harm anyone or act impermissibly by joyriding in a gas-guzzling SUV, this does not mean that our decision is not still open to moral scrutiny. On the contrary, we still might be criticized on the grounds that we have failed to display certain character virtues, such as temperance, moderation, or mindfulness, or that we lack a compelling justification for engaging in such a wasteful hobby. In Rieder's view, the same reasoning may be used to support what he calls a *small family ethic*, wherein procreators see themselves as having significant moral reasons to limit themselves to one or perhaps two children. As he writes,

> The mindful, temperate person, then, who both recognizes and accepts the justificatory burden of procreating, and who exercises restraint and moderation, would plausibly find it difficult to justify having more than one child. After all, once one has had the great fortune to start a family by having a child, adding another child remains exceedingly costly to the environment, but has much less value to the parents; while having a first child takes one from being non-parents to being parents, having a second child merely changes the size of one's family.[44]

This type of approach presents the advantage of supporting a moral prescription in favor of having fewer children while not taking on board an implausibly demanding conception of our climate-related duties. In so doing, however, it also loses some of the teeth associated with arguments that are premised on moral duties and framed in the language of moral permissibility. Procreative limitarians want to claim that we have "a moral obligation to refrain from choosing to have children,"[45] or that "procreation is immoral on environmental grounds,"[46] though supporting these claims via the environmental argument for limiting procreation has proven difficult given the significant challenges associated with defending a plausible and non-arbitrary account of the limits of permissible consumption. Barring the successful defense of such an account, procreative limitarians seem to be left with a version of the argument that is less stringent and perhaps less precise on the question of "how many children is too many?" namely:

1 We have significant moral reasons to limit our individual consumption given its adverse environmental impacts;
2 Being responsible for the consumption of resources as a result of the voluntary decision to procreate is morally equivalent to individually consuming resources; therefore,
3 We have significant moral reasons to limit the number of children we create.

7.5 Are Procreation and Consumption Morally Equivalent?

I have tried to show that, even if we accept the view that procreation and consumption are morally equivalent, the environmental argument for limiting procreation still faces a host of significant challenges in generating moral limits on procreation. We still might wonder, however, whether we should even entertain the idea that procreation and consumption are morally equivalent. Even if they resulted in the same amount of CO_2 emissions, is the creation of a child really morally equivalent to joyriding in a gas-guzzling SUV? Do we really have just as much reason to refrain from the former as we do the latter?

I believe the answer to this question is quite plausibly "no": we have reasons to treat procreation differently from other forms of environmentally impactful activities based on the importance of the interest it makes possible, namely, the interest in establishing a parent-child relationship. It is important to note, however, that rejecting the moral equivalence thesis by affirming the special value of procreation does not give individuals *carte blanche* to procreate indiscriminately, nor is it a problem-free response to the environmental argument for limiting procreation, as it requires us to take a controversial ethical stand with respect to competing conceptions of the good life. I would like to conclude my discussion by briefly considering what rejecting the moral equivalence thesis requires, and what this rejection implies about the morality of procreation.

The most straightforward way to reject the moral equivalence thesis is to show that procreation implicates special interests or values that are not generally implicated by other high-consumption activities. In this book, I have argued that procreation is valuable as the primary and in some cases only means of initiating a parent-child relationship, the participation in which makes a distinctive and non-substitutable contribution to the flourishing of adults. If this account is correct, then it appears that the Grays have a justification for their level of consumption that the Greens lack: while the consumption of the Grays is necessary to satisfy a morally weighty and non-substitutable interest, the consumption of the Greens is not.[47]

But how can we make this claim about the Greens? From what perspective can we designate *their* reasons for consumption as being less weighty or morally important? It cannot be from a *subjective* perspective, as it might be the case that the Greens regard their consumption as being necessary for the satisfaction of what *they* take to be important interests. For example, while they may not see any value in parenting children, they may see significant value in traveling the world, eating gourmet meals, and experiencing the various comforts and luxuries that life has to offer. They may regard their interest in the enjoyment of these goods as non-substitutable, and they may derive just as much happiness from its satisfaction as the Grays derive from the satisfaction of their interest in parenting.[48]

Thus, when we claim that the Grays have a justification for their level of consumption that the Greens lack, we are making this judgment from an *objective* perspective: we are claiming, in effect, that the interest in parenting is objectively more important, and hence has greater justificatory power, than other types of interests. For those who are convinced by the account of the value of parenting drawn upon in this book, this will seem like a plausible position; for those who are not convinced by this account, it may seem objectionably perfectionistic, particularly if it is used to support law or public policy aimed at privileging procreation over other high-consumption activities. In either case, however, those who reject the moral equivalence thesis by appeal to the special value of procreation should be clear about the nature of the claim they are making and the types of objections that it makes them vulnerable to.

Even if this claim is justified, however, what follows from it? Surprisingly little. Even if we are justified in treating procreation differently than other high-consumption activities, it does not then follow that we do not face any moral limits on procreation, or that we are not subject to moral criticism for the creation of multiple children. This is because, as I argued in Chapter 1, the interest that justifies treating procreation differently can be satisfied by the creation of only one child, or even no children if there are suitable children available for adoption—thus, it cannot play a similar justificatory role with respect to the creation of additional children. This does not imply that it is morally wrong to create more than one child, but that our interest in doing so may ultimately be discretionary. In this sense, it may be plausibly regarded as morally equivalent to other forms of non-essential consumption.

Notes

1 Paul A. Murtaugh and Michael G. Schlax. "Reproduction and the Carbon Legacies of Individuals," *Global Environmental Change*, 19 (2009), pp. 14–20.
2 Murtaugh and Schlax, "Reproduction and the Carbon Legacies of Individuals," 18. As I will explain further below in Section 7.3, some of these calculations rely on controversial and ultimately unworkable views about how much of a child's consumption is legitimately attributable to those who are responsible for the existence. The ecological impact of procreation is sometimes grossly overexaggerated on the basis of these views, even if it is still quite significant on more plausible views.
3 Prominent defenses of this type of view include Daniel Burkett. "A Legacy of Harm? Climate Change and the Carbon Cost of Procreation," *Journal of Applied Philosophy*, 38(5) (2021), pp. 790–808; Sarah Conly. *One Child: Do We Have a Right to More?* (Oxford: Oxford University Press, 2015); Trevor Hedberg. *The Environmental Impact of Overpopulation: The Ethics of Procreation* (New York: Routledge, 2020); Corey MacIver. "Procreation or Appropriation?" in Sarah Hannan, Samantha Brennan, and Richard Vernon (eds.) *Permissible Progeny? The Morality of Procreation and Parenting* (Oxford: Oxford University Press, 2015); Christine Overall. *Why Have Children? The Ethical Debate* (Cambridge, MA: MIT Press, 2012), esp. chapter 9; Travis N. Rieder. *Toward a Small Family Ethic: How Overpopulation and Climate Change Are Affecting the Morality of Procreation* (Dordrecht: Springer, 2016); Chad Vance. "Procreation is Immoral on Environmental Grounds," *The Journal of Ethics*, 28(1) (2024), pp. 101–124; and Thomas Young. "Overconsumption and Procreation: Are they Morally Equivalent?" *Journal of Applied Philosophy*, 18(2) (2001), pp. 183–192.
4 Young, "Overconsumption and Procreation," 183.
5 See, for example, Clare Heyward. "A Growing Problem? Dealing with Population Increases in Climate Justice," *Ethical Perspectives*, 19(4) (2012), pp. 703–732, and Ingrid Robeyns. "Is Procreation Special?" *Journal of Value Inquiry*, 56(4) (2022), pp. 643–661.
6 The term "procreative limitarian" was coined by Kalle Grill, based on Ingrid Robeyns' concept of limitarianism in the context of distributive justice. See Kalle Grill. "Procreation vs. Consumption: Harms and Benefits," *Environmental Ethics*, 45(3) (2023), pp. 265–286, 266.
7 For qualified defenses of this strong view, see Young, "Overconsumption and Procreation"; Burkett. "A Legacy of Harm?"; and Vance. "Procreation is Immoral on Environmental Grounds."
8 See, e.g., Conly, *One Child*.
9 See, e.g., Overall, *Why Have Children?* chapter 9.
10 See, e.g., MacIver, "Procreation or Appropriation?".
11 MacIver, "Procreation or Appropriation?" 109.
12 MacIver, "Procreation or Appropriation?" 109–110.
13 MacIver, "Procreation or Appropriation?" 115.

14 Murtaugh and Schlax, "Reproduction and the Carbon Legacy of Individuals."
15 Murtaugh and Schlax, "Reproduction and the Carbon Legacy of Individuals," 14. Note that, on this type of view, genetic relatedness is understood as a proxy for moral responsibility, though as I argued in Chapter 4, these variables come apart: one could be totally morally responsible for the existence of a child to whom one bears no genetic connection, as in the case of *Gamete Thief*.
16 MacIver, "Procreation or Appropriation?" 115.
17 Notice that, on the essential consumption view, a person is not responsible for the totality of their own consumption, for their own essential consumption is attributable to their parents.
18 MacIver, "Procreation or Appropriation?" 117.
19 Some philosophers have suggested that children become *totally* morally responsible for their own consumption once they develop the capacities for agency and moral responsibility, though this view is implausible, as it overlooks the fact that the adults children will become still cannot avoid the level of consumption that is necessary for them to live minimally decent lives, and hence cannot reasonably be held morally responsible for it. For a defense of this type of view, see Phillipe van Basshuysen and Eric Brandstedt. "Comment on 'The Climate Mitigation Gap: Education and Government Recommendations Miss the Most Effective Individual Actions,'" *Environmental Research Letters*, 13(4) (2018), pp. 1–3.
20 This is based on calculations by Felix Pinkert and Martin Sticker, who assume that basic subsistence emissions are determined by the maximally-carbon-frugal lifestyle, which they estimate to be between 1.4 and 4.4 tons of CO_2 per year. As the average North American emits over 16 tons of CO_2 per year, subsistence emissions would account for 25% or less of total emissions. See Felix Pinkert and Martin Sticker, "Procreation, Footprint, and Responsibility for Climate Change," *Journal of Value Inquiry*, 25(3) (2021), pp. 293–321, 305–306.
21 Murtaugh and Schlax, "Reproduction and the Carbon Legacies of Individuals," 18. In addition to its reliance on the unworkable total lifetime consumption view, we might also reject Murtaugh and Schlax's calculation based on the empirical assumption that emissions rates will remain relatively constant. The authors' calculation assume that these rates will remain constant for at least four centuries, though more optimistic scenarios predict carbon neutrality by the mid-21st century. If this were the case, then a child created today would only be a net carbon contributor for a portion of their lifetime, and any descendants they produced would be carbon neutral; thus, their contribution to the carbon legacy of their parents would be much smaller *even on the total lifetime consumption view*. For further discussion, see Felix Pinkert and Martin Sticker. "Procreative Prerogatives and Climate Change," *Journal of Applied Philosophy*, 5–6, https://doi.org/10.1111/japp.12773.
22 This is based on the assumption that subsistence emissions account for around 25% of the 16 tons of CO_2 emitted annually by the average North American, leaving 12 tons of CO_2 per year in non-essential emissions, or 216 tons of CO_2 in non-essential emissions during the first 18 years of a child's life.
23 This is based on figures presented by Seth Wynes and Kimberley A. Nichols, according to which living car-free reduces one's carbon impact by 2.4 tons of CO_2 per year, while adopting a meatless diet reduces one's carbon impact by 0.9 tons of CO_2 per year. See Seth Wynes and Kimberley A. Nicholas. "The Climate Mitigation Gap: Education and Government Recommendations Miss the Most Effective Individual Actions," *Environmental Research Letters* 12 (2017), pp. 1–9, 4.

24 MacIver, "Procreation or Appropriation?", 117. Emphasis added.
25 Walter Sinnott-Armstrong. "It's Not My Fault: Global Warming and Individual Moral Obligations," in Stephen M. Gardiner, Simon Caney, Dale Jamieson, and Henry Shue (eds.) *Climate Ethics: Essential Readings* (New York: Oxford University Press, 2010), pp. 332–346.
26 As many commentators have pointed out, this view faces the obvious objection that our political support is also causally inefficacious and hence not morally required (e.g., if our vote will not make a difference to the outcome of an election). See, for example, Avram Hiller. "Climate Change and Individual Responsibility," *The Monist*, 94(3) (2011), pp. 349–368, esp. 364–365.
27 Amartya Sen is a proponent of this view. See, for example, "Fertility and Coercion," *University of Chicago Law Review*, 63(3) (1996), pp. 1035–1061.
28 Sinnott-Armstrong, "It's Not *My* Fault," 298–299.
29 Vance, "Procreation is Immoral on Environmental Grounds," 117.
30 John Nolt. "How Harmful Are the Average American's Greenhouse Gas Emissions?" *Ethics, Policy, and Environment*, 14(1) (2011), pp. 3–10, 9.
31 Nolt, "How Harmful are the Average American's Greenhouse Gas Emissions?" 9.
32 For a recent exploration of this worry and different approaches toward addressing it, see Chad Vance. "Justifying Subsistence Emissions: An Appeal to Causal Impotence," *Journal of Value Inquiry*, 57(3) (2021), pp. 515–532.
33 This could be either directly *qua* human beings or indirectly *qua* citizens of states, depending among whom one thinks permissions to consume resources ought to be fairly distributed. I will assume the former view in this chapter.
34 There is an obvious objection to this claim based on the non-identity problem, namely, that insofar as our consumption decisions are identity-affecting, we will not be making future people any worse off by failing to discharge our sustainability-based duties so long as they have lives worth living. The rights-based solution to the non-identity problem defended in Chapter 2 may be inserted here in response, as it has the resources to explain how future people may be wronged by our failure to fulfill our sustainability-based duties even if they are not made worse off.
35 For a similar "fair shares" approach to understanding the wrongness of individual overconsumption, see Christian Baatz. "Climate Change and Individual Duties to Reduce GHG Emissions," *Ethics, Policy & Environment*, 17(1) (2014), pp. 1–19.
36 Simon Caney. "Human Rights, Population, and Climate Change," in Dapo Akande, Jaakko Kuosmanen, Helen McDermott, and Dominic Roser (eds.) *Human Rights and 21st Century Challenges* (Oxford: Oxford University Press, 2020).
37 Conly defends a specific limit of one child per couple, whereas Overall defends a specific limit of one child per person (or two children per couple). See Conly, *One Child, and Overall, Why Have Children?* esp. chapter 9.
38 Intergovernmental Panel on Climate Change (IPCC). *Climate Change: 2023 Synthesis Report* (Geneva: IPCC, 2023), 98.
39 Samuel Scheffler. *Death and the Afterlife*. Ed. Niko Kolodny (Oxford: Oxford University Press, 2013).
40 Conly, *One Child*, esp. chapter 2.
41 Tim Meijers. "Climate Change and the Right to One Child," in Gerhard Bos and Marcus Düwell (eds.) *Human Rights and Sustainability: Moral Responsibilities for the Future* (New York: Routledge, 2016), 188.

42 For a similar argument, see Elizabeth Cripps. "Why It Can Be Permissible to Have Kids in the Climate Emergency," *Journal of Applied Philosophy*, https://doi.org/10.1111/japp.12756.
43 Rieder, *Toward a Small Family Ethic.*
44 Rieder, *Toward a Small Family Ethic*, 58.
45 Burkett, "A Legacy of Harm?" 791.
46 Vance, "Procreation Is Immoral on Environmental Grounds," 103.
47 Pinker and Sticker defend a similar view in "Procreative Prerogatives and Climate Change."
48 They may even derive *more* happiness from the satisfaction of this interest, as empirical evidence suggests that parenting does not in fact increase happiness. For an account of this evidence, see Thomas Hansen. "Parenthood and Happiness: A Review of Folk Theories versus Empirical Evidence," *Social Indicators Research*, 108(1) (2012), pp. 1–36. For a critical discussion, see Cripps, "Why It Can Be Permissible to Have Kids," esp. 4–7.

Bibliography

Almond, Brenda. "Family Relationships and Reproductive Technology," In *Having and Raising Children: Unconventional Families, Hard Choices, and the Social Good*, edited by U. Narayan & J.J. Bartkowiak. University Park, PA: Pennsylvania University Press, 1999.

Alstott, Anne L. *No Exit: What Parents Owe Their Children and What Society Owes Parents*. Oxford: Oxford University Press, 2004.

Archard, David. *Children: Rights and Childhood (second edition)*. New York: Routledge, 2004a.

Archard, David. "Wrongful Life," *Philosophy*, 79(309) (2004b), pp. 403–420.

Archard, David. "The Obligations and Responsibilities of Parenthood," In *Procreation and Parenthood: The Ethics of Bearing and Rearing Children*, edited by David Archard and David Benatar. Oxford: Oxford University Press, 2010a.

Archard, David. *The Family: A Liberal Defense*. New York: Palgrave MacMillan, 2010b.

Arneson, Richard J. "The Principle of Fairness and Free-Rider Problems," *Ethics*, 92(4) (1982), pp. 616–633.

Austin, Michael W. *Conceptions of Parenthood: Ethics and the Family*. Burlington: Ashgate, 2007.

Baatz, Christian. "Climate Change and Individual Duties to Reduce GHG Emissions," *Ethics, Policy & Environment*, 17(1) (2014), pp. 1–19.

Baron, Teresa. "A Lost Cause? Fundamental Problems for Causal Theories of Parenthood," *Bioethics*, 34(7) (2020), pp. 664–670.

Bartholet, Elizabeth. *Family Bonds: Adoption, Infertility, and the New World of Child Production*. Boston: Beacon Press, 1999a.

Bartholet, Elizabeth. *Nobody's Children: Abuse and Neglect, Foster Drift, and the Adoption Alternative*. Boston: Beacon Press, 1999b.

Bartholet, Elizabeth. "International Adoption: The Human Rights Position," *Global Policy*, 1(1) (2010), pp. 91–100.

Bayles, Michael. "Harm to the Unconceived," *Philosophy and Public Affairs*, 5(3) (1976), pp. 292–304.

Bayne, Tim. "In Defense of Genethical Parity," In *Procreation and Parenthood: The Ethics of Bearing and Rearing Children*, edited by David Archard and David Benatar. Oxford: Oxford University Press, 2010.

Bayne, Tim and Avery Kolers. "'Are you My Mommy?' On the Genetic Basis of Parenthood," *Journal of Applied Philosophy*, 18(3) (2001), pp. 273–285.

Bayne, Tim and Avery Kolers. "Towards a Pluralist Account of Parenthood," *Bioethics*, 17(3) (2003), pp. 221–242.

Beckerman, Wilfred and Joanna Pasek. *Justice, Posterity, and the Environment: Environmental Ethics for a New Millennium.* Oxford: Oxford University Press, 2001.

Begle, Angela Moreland, Jean E. Dumas, and Rochelle F. Hanson. "Predicting Child Abuse Potential: An Empirical Investigation of Two Theoretical Frameworks," *Journal of Clinical Child & Adolescent Psychology*, 39(2) (2010), pp. 208–219.

Benatar, David. "The Unbearable Lightness of Bringing into Being," *Journal of Applied Philosophy*, 16(2) (1999), pp. 173–180.

Benatar, David. *Better Never to Have Been: The Harm of Coming into Existence.* Oxford: Oxford University Press, 2006.

Benatar, David. "Famine, Affluence, and Procreation: Peter Singer and Anti-Natalism Lite," *Ethical Theory and Moral Practice*, 23(2) (2020), pp. 415–431.

Benatar, David and David Wasserman. *Debating Procreation: Is it Wrong to Reproduce?* Oxford: Oxford University Press, 2015.

Blustein, Jeffrey. *Parents and Children: The Ethics of the Family.* New York: Oxford University Press, 1982.

Blustein, Jeffrey. "Procreation and Parental Responsibility," *Journal of Social Philosophy*, 28(2) (1997), pp. 79–86.

Boonin, David. *The Non-Identity Problem and the Ethics of Future People.* Oxford: Oxford University Press, 2014.

Botterell, Andrew and Carolyn MacLeod. "'Not for the Faint of Heart': Assessing the Status Quo on Parental Licensing," In *Family Making: Contemporary Ethical Challenges*, edited by Carolyn McLeod and Françoise Baylis. Oxford: Oxford University Press, 2014.

Bou-Habib, Paul. "Parental Subsidies: The Argument from Insurance," *Politics, Philosophy, and Economics*, 12(2), pp. 197–216.

Brake, Elizabeth. "Fatherhood and Child Support: Do Men Have a Right to Choose?" *Journal of Applied Philosophy*, 22(1) (2005), pp. 55–73.

Brake, Elizabeth. "Willing Parents: A Voluntarist Account of Parental Role Obligations," In *Procreation and Parenthood: The Ethics of Bearing and Rearing Children*, edited by David Archard and David Benatar. Oxford: Oxford University Press, 2010.

Brake, Elizabeth. "Creation Theory: Do Genetic Ties Matter?" In *Permissible Progeny? The Morality of Procreation and Parenting*, edited by Sarah Hannan, Samantha Brennan, and Richard Vernon. New York: Oxford University Press, 2015.

Brandt, Reuven. "The Transfer and Delegation of Responsibilities for Genetic Offspring in Gamete Donation," *Journal of Applied Philosophy*, 34(5) (2017), pp. 665–678.

Brennan, Samantha. "The Goods of Childhood and Children's Rights," In *Family-Making: Contemporary Ethical Challenges*, edited by Françoise Baylis and Carolyn McLeod. Oxford: Oxford University Press, 2014.

Brennan, Samantha and Robert Noggle. "The Moral Status of Children: Children's Rights, Parents' Rights, and Family Justice," *Social Theory and Practice*, 23(1) (1997), pp. 1–26.

Brighouse, Harry and Adam Swift. "Parents' Rights and the Value of the Family," *Ethics*, 117(1) (2006), pp. 80–108.

Brighouse, Harry and Adam Swift. *Family Values: The Ethics of Parent-Child Relationships*. Princeton: Princeton University Press, 2014.

Brock, Dan W. "Genetic Harms and the Non-Identity Problem: The Case of Wrongful Handicaps," *Bioethics*, 9(3/4) (1995), pp. 269–275.

Brock, Dan W. "Shaping Future Children: Parental Rights and Societal Interests," *Journal of Political Philosophy*, 13(4) (2005), pp. 377–398.

Buchanan, Allen, Dan W. Brock, Norman Daniels, and Daniel Wikler. *From Chance to Choice: Genetics and Justice*. Cambridge: Cambridge University Press, 2000.

Bucharest Early Intervention Project. (2023) "About the Bucharest Early Intervention Project," accessed July 2023 from https://www.bucharestearlyinterventionproject.org/

Burkett, Daniel. "A Legacy of Harm? Climate Change and the Carbon Cost of Procreation," *Journal of Applied Philosophy*, 38(5) (2021), pp. 790–808.

Burley, Justine. "The Price of Eggs: Who Should Bear the Costs of Fertility Treatment?" In *The Future of Human Reproduction*, edited by John Harris and Soren Holm Oxford: Oxford University Press, 1998.

Callahan, Daniel. "Bioethics and Fatherhood," *Utah Law Review*, 3 (1992), pp. 735–746.

Caney, Simon. "Human Rights, Population, and Climate Change," In *Human Rights and 21st Century Challenges*, edited by Dapo Akande, Jaakko Kuosmanen, Helen McDermott, and Dominic Roser Oxford: Oxford University Press, 2020.

Casal, Paula. "Environmentalism, Procreation, and the Principle of Fairness," *Public Affairs Quarterly*, 13(4) (1999), pp. 363–376.

Casal, Paula and Andrew Williams. "Equality of Resources and Procreative Justice," In *Dworkin and His Critics*, edited by Justine Burley. Malden, MA.: Blackwell, 2004.

Child Welfare Information Gateway. *Costs of Adopting: A Factsheet for Families*. Washington, DC: U.S. Department of Health and Human Services, 2011.

Clayton, Matthew. *Justice and Legitimacy in Upbringing*. Oxford: Oxford University Press, 2006.

Cohen, Cynthia B. "The Morality of Knowingly Conceiving Children with Serious Conditions: An Expanded 'Wrongful Life' Standard," In *Contingent Future Persons: On the Ethics of Deciding Who Will Live, or Not, in the Future*, edited by Nick Fotion and Jan C. Heller. Dordrecht: Springer, 1997.

Cohen, G.A. "On the Currency of Egalitarian Justice," *Ethics*, 99(4) (1989), pp. 906–944.

Conly, Sarah. *One Child: Do we have a Right to More?* Oxford: Oxford University Press, 2015.

Council of Europe. *European Convention on Human Rights* 3 September 1953.

Cripps, Elizabeth. "Why It Can Be Permissible to Have Kids in the Climate Emergency," *Journal of Applied Philosophy*, 42 (2025), pp. 24–43. https://doi.org/10.1111/japp.12756

Cutas, Daniela. "Sex Is Overrated: On the Right to Reproduce," *Human Fertility*, 12(1) (2009), pp. 45–52.

Daly, Martin and Margo Wilson. *The Truth about Cinderella: A Darwinian View of Parental Love*. New Haven, CT: Yale University Press, 1999.

De George, Richard. "The Environment, Rights, and Future Generations," In *Ethics and Problems of the 21st Century*, edited by Kenneth Goodpaster and K.M. Sayre Notre Dame: Notre Dame University Press, 1979.

De Melo Martin, Inmaculada. "The Ethics of Anonymous Gamete Donation: Is There a Right to Know One's Genetic Origins?" *Hasting Center Report*, 44(2) (2014), pp. 28–35.

De Wispelaere, Jurgen and Daniel Weinstock. "State Regulation and Assisted Reproduction: Balancing the Interests of Parents and Children," In *Family-Making:*

Contemporary Ethical Challenges, edited by François Baylis and Carolyn McLeod. Oxford: Oxford University Press, 2014.

De Wispelaere, Jurgen and Daniel Weinstock. "Privileging Adoption Over Sexual Reproduction? A State-Centered Perspective," In *Permissible Progeny? The Morality of Procreation and Parenting*, edited by Sarah Hannan, Samantha Brennan, and Richard Vernon. New York: Oxford University Press, 2015.

DeGrazia, David. *Creation Ethics: Reproduction, Genetics, and Quality of Life*. Oxford: Oxford University Press, 2012.

Durà-Vilà, Victor. "Parental Obligation, Adoption, and Abortion: Critique of Porter and a Nozickean Alternative," *Journal of Value Inquiry*, 47(1–2) (2013), pp. 29–47.

Dworkin, Ronald. "Rights as Trumps," In *Theories of Rights*, edited by Jeremy Waldron. (Oxford: Oxford University Press, 1984).

Dworkin, Ronald. *Sovereign Virtue: The Theory and Practice of Equality*. Cambridge, MA: Harvard University Press, 2000.

Elliott, Robert. "The Rights of Future People," *Journal of Applied Philosophy*, 6(2) (1989), pp. 59–69.

Feinberg, Joel. *Harm to Others, Vol. 1 of the Moral Limits of the Criminal Law*. New York: Oxford University Press, 1984.

Feinberg, Joel. "Wrongful Life and the Counterfactual Element in Harming," In *Freedom and Fulfillment: Philosophical Essays*. Princeton: Princeton University Press, 1992.

Ferracioli, Luara. "Procreative-Parenting, Love's Reasons, and the Demands of Morality," *Philosophical Quarterly*, 68(270) (2018), pp. 77–97.

Ferracioli, Luara. (n.d.) "On the Human Right to Found a Family," In *The Routledge Handbook for the Philosophy of Human Rights*, edited by Jesse Tomalty and Kerri Woods. New York: Routledge, forthcoming.

Folbre, Nancy. *The Invisible Heart*. New York: New Press, 2001.

Friedrich, Daniel. "A Duty to Adopt?" *Journal of Applied Philosophy*, 30(1) (2013), pp. 25–39.

Fuscaldo, Giuliana. "Genetic Ties: Are They Morally Binding?" *Bioethics*, 20(2) (2006), pp. 64–76.

George, Rolf. "Who Should Bear the Costs of Children?" *Public Affairs Quarterly*, 1(1) (1987), pp. 1–42.

Gerhardt, Sue. *Why Love Matters: How Affection Shapes a Baby's Brain*. New York: Routledge, 2004.

Gheaus, Anca. "Arguments for Nonparental Care for Children," *Social Theory and Practice*, 37(3) (2011), pp. 483–509.

Gheaus, Anca. "The Right to Parent One's Biological Baby," *Journal of Political Philosophy*, 20(4) (2012), pp. 432–455.

Gheaus, Anca. "Unfinished Adults and Defective Children: On the Nature and Value of Childhood," *Journal of Ethics and Social Philosophy*, 9(1) (2015), pp. 1–21.

Gheaus, Anca. "Children's Vulnerability and Legitimate Authority over Children," *Journal of Applied Philosophy*, 35(S1) (2017), pp. 60–75.

Gheaus, Anca. "Biological Parenthood: Gestational, Not Genetic," *Australasian Journal of Philosophy*, 96(2) (2018), pp. 225–240.

Gheaus, Anca. "More Co-Parents, Fewer Children: Multiparenting and Sustainable Population," *Essays in Philosophy*, 20(1) (2019), pp. 1–21.

Gheaus, Anca. "The Best Available Parent," *Ethics*, 131(3) (2021), pp. 431–459.

Goldfarb, William. "Effects of Psychological Deprivation in Infancy and Subsequent Stimulation," *American Journal of Psychology*, 102(1) (1945), pp. 18–33.

Graff, E.J. "The Lie We Love," *Foreign Policy*, 169 (2008), pp. 58–66.

Grill, Kalle. "Procreation vs. Consumption: Harms and Benefits," *Environmental Ethics*, 45(3) (2023), pp. 265–286.

Hales, Stephen D. "Abortion and Father's Rights," In *Reproduction, Technology, and Rights*, edited by James M Humber and Robert F. Almeder. Totowa: Humana Press, 1996.

Hall, Barbara. "The Origin of Parental Rights," *Public Affairs Quarterly*, 13(1) (1999), pp. 73–82.

Hanna, Jason. "Causal Parenthood and the Ethics of Gamete Donation," *Bioethics*, 33(2) (2019), pp. 267–273.

Hannan, Sarah. "Introduction," In *Permissible Progeny? The Morality of Procreation and Parenting*, edited by Sarah Hannan, Samantha Brennan, and Richard Vernon. Oxford: Oxford University Press, 2015.

Hannan, Sarah. "Why Childhood is Bad for Children," *Journal of Applied Philosophy*, 35(S1) (2018), pp. 11–28.

Hannan, Sarah and R.J. Leland. "Childhood Bads, Parenting Goods, and the Right to Procreate," *Critical Review of International Social and Political Philosophy*, 21(3) (2018), pp. 366–384.

Hannan, Sarah and Richard Vernon. "Parental Rights: A Role-Based Approach," *Theory and Research in Education*, 6(2) (2008), pp. 173–189.

Hansen, T. "Parenthood and Happiness: A Review of Folk Theories versus Empirical Evidence," *Social Indicators Research*, 108(1) (2012), pp. 29–64.

Hardimon, Michael. "Role Obligations," *Journal of Philosophy*, 91(7) (1994), pp. 333–363.

Harman, Elizabeth. "Can We Harm and Benefit in Creating?" *Philosophical Perspectives*, 18 (2004), pp. 89–113.

Harris, John. "The Right to Found a Family," In *Children, Parents, and Politics*, edited by Geoffrey Scarre. Cambridge: Cambridge University Press, 1989.

Hart, H.L.A. "Are There Any Natural Rights?" In *Theories of Rights*, edited by Jeremy Waldron. Oxford: Oxford University Press, 1984.

Haslanger, Sally. *Resisting Reality: Social Construction and Social Critique*. Oxford: Oxford University Press, 2012.

Häyry, Matti. "A Rational Cure for Pre-reproductive Stress Syndrome," *Journal of Medical Ethics*, 30(4) (2004), pp. 377–378.

Hedberg, Trevor. *The Environmental Impact of Overpopulation: The Ethics of Procreation*. New York: Routledge, 2020.

Heyd, David. *Genethics: Moral Issues in the Creation of People*. Berkeley: University of California Press, 1992.

Heyd, David. "The Intractability of the Nonidentity Problem," In *Harming Future Persons: Ethics, Genetics, and the Nonidentity Problem*, edited by Melinda A. Roberts and David T. Wasserman. Dordrecht: Springer, 2009.

Heyward, Clare. "A Growing Problem? Dealing with Population Increases in Climate Justice," *Ethical Perspectives*, 19(4) (2012), pp. 703–732.

Hohfeld, Wesley. *Fundamental Legal Conceptions*, edited by W.W. Cook. New Haven: Yale University Press, 1919.

Holtug, Nils. "On the Value of Coming into Existence," *Journal of Ethics*, 5(4) (2001), pp. 361–384.

Honoré, Tony. *Responsibility and Fault*. Oxford: Hart Publishing, 1999.

Hubin, Don. "Procreators' Duties: Sexual Asymmetries," In *The Oxford Handbook of Reproductive Ethics*, edited by Leslie Francis. Oxford: Oxford University Press, 2017.

Hurley, Paul and Rivka Weinberg. "Whose Problem is Non-Identity?" *Journal of Moral Philosophy*, 12(6) (2015), pp. 699–730.

Intergovernmental Panel on Climate Change (IPCC). *Climate Change: 2023 Synthesis Report* Geneva: IPCC, 2023.

Jeske, Diane. "Families, Friends, and Special Obligations," *Canadian Journal of Philosophy*, 28(4) (1998), pp. 527–556.

Joyce, Kathryn. *The Child Catchers: Rescue, Trafficking, and the New Gospel of Adoption*. New York: PublicAffairs, 2013.

Kavka, Gregory S. "The Paradox of Future Individuals," *Philosophy and Public Affairs*, 11(2) (1982), pp. 93–112.

Klosko, George. "Presumptive Benefit, Fairness, and Political Obligation," *Philosophy and Public Affairs*, 16(3) (1987), pp. 241–259.

Koepke, J.E., S. Anglin, J. Austin, and J. Delesalle. "Becoming Parents: Feelings of Adoptive Mothers," *Pediatric Nursing*, 17(4) (1991), pp. 333–336.

Kramer, Matthew, N.E. Simmonds, and Hillel Steiner. *A Debate over Rights: Philosophical Enquiries*. Oxford: Oxford University Press, 1998.

Kumar, Rahul. "Who Can Be Wronged?" *Philosophy and Public Affairs*, 31(2) (2003), pp. 99–118.

Levy, Neil and Mianna Lotz. "Reproductive Cloning and a (Kind of) Genetic Fallacy," *Bioethics*, 19(3) (2005), pp. 232–250.

Lewis, David. "Causation," *Journal of Philosophy*, 70(17) (1973), pp. 556–567.

Liao, S. Matthew. *The Right to Be Loved*. Oxford: Oxford University Press, 2015.

Lino, M., K. Kuczynski, N. Rodriguez, and T. Schap. *Expenditures on Children by Families, 2015*. Washington, DC: U.S. Department of Agriculture, 2017.

Locke, John. *Second Treatise of Government*, edited by C.B. Macpherson. Indianapolis: Hackett, 1980.

MacCormick, Neil. "Rights in Legislation," In *Law, Morality, and Society: Essays in Honour of H.L.A. Hart*, edited by P.M.S. Hacker and Joseph Raz. Oxford: Clarendon Press, 1977.

MacIver, Corey. "Procreation or Appropriation?" In *Permissibly Progeny? The Morality of Procreation and Parenting*, edited by Sarah Hannan, Samantha Brennan, and Richard Vernon. New York: Oxford University Press, 2015.

Macklin, Ruth. "Can Future Generations Correctly be said to Have Rights?" In *Responsibilities to Future Generations*, edited by Ernest Partridge. Buffalo: Prometheus Books, 1981.

MacLeod, Colin. "Parental Responsibilities in an Unjust World," In *Procreation and Parenthood: The Ethics of Bearing and Rearing Children*, edited by David Archard and David Benatar. Oxford: Oxford University Press, 2010.

MacLeod, Colin. "Parental Competency and the Right to Parent," In*Permissibly Progeny? The Morality of Procreation and Parenting*, edited by Sarah Hannan, Samantha Brennan, and Richard Vernon. New York: Oxford University Press, 2015.

Magnusson, Erik. "Parental Justice and the Kids Pay View," *Ethical Theory and Moral Practice*, 21(4) (2018), pp. 963–977.

Magnusson, Erik. "How to Reject Benatar's Asymmetry Argument," *Bioethics*, 33(6) (2019), pp. 674–683.

Magnusson, Erik. "On Risk-Based Arguments for Anti-Natalism," *Journal of Value Inquiry*, 56(1) (2022), pp. 101–117.

Marsh, Jason. "Quality of Life Assessments, Cognitive Reliability, and Procreative Responsibility," *Philosophy and Phenomenological Research*, 89(2) (2014), pp. 436–466.

McMahan, Jeff. *The Ethics of Killing: Problems at the Margins of Life*. Oxford: Oxford University Press, 2002.

McTernan, Emily. "Should Fertility Treatment Be State Funded?" *Journal of Applied Philosophy*, 32(3) (2015), pp. 227–240.

Meijers, Tim. "Climate Change and the Right to One Child," In *Human Rights and Sustainability: Moral Responsibilities for the Future*, edited by Gerhard Bos and Marcus Düwell. New York: Routledge, 2016.

Meijers, Tim. "The Value in Procreation: A Pro-Tanto Case for a Limited and Conditional Right to Procreate," *Journal of Value Inquiry*, 54(4) (2020), pp. 627–647.

Miller, David. "Distributing Responsibilities," *Journal of Political Philosophy*, 9(4) (2001), pp. 453–471.

Millum, Joseph. "How Do We Acquire Parental Responsibilities?" *Social Theory and Practice*, 34(1) (2008), pp. 71–93.

Millum, Joseph. *The Moral Foundations of Parenthood*. Oxford: Oxford University Press, 2016.

Moschella, Melissa. "Rethinking the Moral Permissibility of Gamete Donation," *Theoretical Medicine and Bioethics*, 35(6) (2014), pp. 421–440.

Mullin, Amy. *Reconceiving Pregnancy and Childcare: Ethics, Experience, and Reproductive Labor*. New York: Cambridge University Press, 2005.

Murphy, Liam B. "The Demands of Beneficence," *Philosophy and Public Affairs*, 22(4) (1993), pp. 267–292.

Murtaugh, Paul A. and Michael G. Schlax. "Reproduction and the Carbon Legacies of Individuals," *Global Environmental Change*, 19 (2009), pp. 14–20.

National Conference of State Legislatures. (2024) "Parental Rights and Sexual Assault," accessed August 2024 from https://www.ncsl.org/human-services/parental-rights-and-sexual-assault

Ndegwa, S., and S. Kelly. "Status of Public Funding for In Vitro Fertilization in Canada and Internationally," *Environmental Scan*, 14 (2010), pp. 1–12.

Nelson, James Lindemann. "Parental Obligations and Surrogacy Contracts: A Causal Perspective," *Public Affairs Quarterly*, 5(1) (1991), pp. 49–61.

Noggle, Robert. "Special Agents: Children's Autonomy and Parental Authority," In *The Moral and Political Status of Children*, edited by David Archard and Colin Macleod. Oxford: Oxford University Press, 2002.

Nolt, John. "How Harmful are the Average American's Greenhouse Gas Emissions?" *Ethics, Policy, and Environment*, 14(1) (2011), pp. 3–10.

Nozick, Robert. *Anarchy, State, and Utopia*. New York: Basic Books, 1974.

O'Neill, Onora. "Begetting, Bearing, and Rearing," In *Having Children: Philosophical and Legal Reflections on Parenthood*, edited by Onora O'Neill and William Ruddick. New York: Oxford University Press, 1979.

Okin, Susan Moller. *Justice, Gender, and the Family*. New York: Basic Books, 1989.

Olsaretti, Serena. "Responsibility and the Consequences of Choice," *Proceedings of the Aristotelian Society*, 109 (2009a), pp. 165–188.

Olsaretti, Serena. "Choice, Circumstance, and the Cost of Children," In *Hillel Steiner and the Anatomy of Justice: Themes and Challenges*, edited by Steven de Wijze, Matthew H. Kramer, and Ian Carter. New York: Routledge, 2009b.

Olsaretti, Serena. "Children as Public Goods?" *Philosophy and Public Affairs*, 41(3) (2013), pp. 226–258.

Olsaretti, Serena. "Liberal Equality and the Moral Status of Parent-Child Relationships," In *Oxford Studies in Political Philosophy*, edited by David

Sobel, Peter Vallentyne, and Steven Wall. Vol. 3. Oxford: Oxford University Press, 2017.

Overall, Christine. *Why Have Children? The Ethical Debate*. Cambridge, MA: MIT Press, 2012.

Parfit, Derek. *Reasons and Persons*. Oxford: Clarendon Press, 1984.

Parfit, Derek. "Comments," *Ethics*, 96(4) (1986), pp. 832–872.

Pearson, Yvette E. "Storks, Cabbage Patches, and the Right to Procreate," *Bioethical Inquiry*, 4 (2007), pp. 105–115.

Petersen, Thomas Søbirk. "The Claim from Adoption," *Bioethics*, 16(4) (2002), pp. 353–374.

Pinkert, Felix and Martin Sticker. "Procreation, Footprint, and Responsibility for Climate Change," *Journal of Value Inquiry*, 25(3) (2021), pp. 293–321.

Pinkert, Felix and Martin Sticker. "Procreative Prerogatives and Climate Change," *Journal of Applied Philosophy*, 42 (2025) 44–66. https://doi.org/10.1111/japp.12773

Plato. *Republic*. Translated by G.M.A. Grube, revised by C.D.C. Reeve. Indianapolis: Hackett, 1992.

Porter, Lindsey. "Adoption is Not Abortion-Lite," *Journal of Applied Philosophy*, 29(1) (2012), pp. 63–76.

Porter, Lindsey. "Why and How to Prefer a Causal Account of Parenthood," *Journal of Social Philosophy*, 45(2) (2014), pp. 182–202.

Porter, Lindsey. "Gestation and Parental Rights: Why is Good Enough Good Enough?" *Feminist Philosophical Quarterly*, 1(1) (2015), pp. 1–27.

Prusak, Bernard. "The Cost of Procreation," *Journal of Social Philosophy*, 42(1) (2011), pp. 61–75.

Prusak, Bernard. *Parental Obligations and Bioethics: The Duties of a Creator*. New York: Routledge, 2013.

Quigley, Muireann. "A Right to Reproduce?" *Bioethics*, 24(8) (2010), pp. 403–411.

Rachels, Stuart. "The Immorality of Having Children," *Ethical Theory and Moral Practice*, 17(3) (2014), pp. 567–582,

Rakowski, Eric. *Equal Justice*. Oxford: Oxford University Press, 1991.

Rape, Abuse, and Incest National Network (RAINN). (2024) "Reporting Rates," accessed October 2024 from https://rainn.org/getinformation/statistics/reporting-rates

Rawls, John. *A Theory of Justice*. Cambridge, MA: Harvard University Press, 1971.

Raz, Joseph. *The Morality of Freedom*. Oxford: Clarendon Press, 1986.

Reiman, Jeffrey. "Being Fair to Future People: The Non-Identity Problem in the Original Position," *Philosophy and Public Affairs*, 35(1) (2007), pp.69–92.

Richards, Norvin. *The Ethics of Parenthood*. Oxford: Oxford University Press, 2010.

Richards, Norvin. "How We Acquire Parental Rights," In *The Oxford Handbook of Reproductive Ethics*, edited by Leslie Francis. (Oxford: Oxford University Press, 2016).

Rieder, Travis N. "Procreation, Adoption, and the Contours of Obligation," *Journal of Applied Philosophy*, 32(3) (2015), pp. 293–309.

Rieder, Travis N. *Toward a Small Family Ethic: How Overpopulation and Climate Change are Affecting the Morality of Procreation*. Dordrecht: Springer, 2016.

Rivera-López, Eduardo. "The Claim from Adoption Revisited," *Bioethics*, 20(6) (2006), pp. 319–325.

Roberts, Melinda. *Child versus Childmaker: Future Persons and Present Duties in Ethics and the Law*. Oxford: Rowman and Littlefield, 1998.

Robertson, John A. *Children of Choice: Freedom and the New Reproductive Technologies*. Princeton: Princeton University Press, 1994.

Robeyns, Ingrid. "Is Procreation Special?" *Journal of Value Inquiry*, 56(4) (2022), pp. 643–661.

Rothman, Barbara Katz. *Reconceiving Motherhood: Ideology and Technology in a Patriarchal Society*. New York: W.W. Norton, 1989.

Rudy-Hiller, Fernando. "The Epistemic Condition for Moral Responsibility," In *Stanford Encyclopedia of Philosophy*, (Winter 2022 Edition), edited by Edward N. Zalta and Uri Nodelman, accessed from https://plato.stanford.edu/archives/win2022/entries/moral-responsibility-epistemic

Rulli, Tina. "The Duty to Adopt," PhD Diss., Yale University, 2012.

Rulli, Tina. "The Unique Value of Adoption," In *Family-Making: Contemporary Ethical Challenges*, edited by Carolyn McLeod and Françoise Baylis. Oxford: Oxford University Press, 2014.

Rulli, Tina. "Preferring a Genetically-Related Child," *Journal of Moral Philosophy*, 13(6) (2016a), pp. 669–698.

Rulli, Tina. "The Ethics of Procreation and Adoption," *Philosophy Compass*, 11(6) (2016b), pp. 305–316.

Sartorio, Carolina. "How to Be Responsible for Something Without Causing It," *Philosophical Perspectives*, 18 (2004), pp. 315–336.

Savulescu, Julian. "Procreative Beneficence: Why We Should Select the Best Children," *Bioethics*, 15(5/6) (2001), pp. 413–426.

Scheffler, Samuel. *The Rejection of Consequentialism*. Oxford: Clarendon Press, 1994.

Scheffler, Samuel. *Death and the Afterlife*, edited by Niko Kolodny. Oxford: Oxford University Press, 2013.

Schoeman, Ferdinand. "Rights of Children, Rights of Parents, and the Moral Basis of the Family," *Ethics*, 91(1) (1980), pp.6–19.

Schwartz, Thomas. "Obligations to Posterity," In *Obligations to Future Generations*, edited by Brian Barry and R.I. Sikora. Philadelphia: Temple University Press, 1978.

Sen, Amartya. "Fertility and Coercion," *University of Chicago Law Review*, 63(3) (1996), pp. 1035–1061.

Shanner, Laura. "The Right to Procreate: Where Rights Claims Have Gone Wrong," *McGill Law Journal*, 40 (1995), pp. 823–874.

Shiffrin, Seana Valentine. "Wrongful Life, Procreative Responsibility, and the Significance of Harm," *Legal Theory*, 5 (1999), pp. 117–148.

Shiffrin, Seana Valentine. "Harm and its Moral Significance," *Legal Theory*, 18 (2012), pp. 1–42.

Shpall, Sam. "Parental Love and Procreation," *Philosophical Quarterly*, 73(1) (2023), pp. 206–226.

Simmons, A. John. "The Principle of Fair Play," *Philosophy and Public Affairs*, 8(4) (1979) pp. 307–337.

Singer, Leslie M., David M. Brodzinsky, Douglas Ramsay, Mary Steir, and Everett Waters. "Mother-Infant Attachment in Adoptive Families," *Child Development*, 56(6) (1985), pp. 1543–1551.

Singer, Peter. "Famine, Affluence, and Morality," *Philosophy and Public Affairs*, 1(3) (1972), pp. 229–243.

Sinnott-Armstrong, Walter. "It's Not My Fault: Global Warming and Individual Moral Obligations," In *Climate Ethics: Essential Readings*, edited by Stephen M. Gardiner, Simon Caney, Dale Jamieson, and Henry Shue. New York: Oxford University Press, 2010.

Smolkin, Doran. "Toward a Rights-Based Solution to the Non-Identity Problem," *Journal of Social Philosophy*, 30(1) (1999), pp. 194–208.

Sommers, Christina Hoff. "Filial Morality," *The Journal of Philosophy*, 83(8) (1986), pp. 439–456.

Statman, Daniel. "The Right to Parenthood: An Argument for a Narrow Interpretation," *Ethical Perspectives*, 10(3–4) (2003), pp. 224–235.

Steinbock, Bonnie. "The Logical Case for 'Wrongful Life'," *Hastings Center Report*, 16(2) (1986), pp. 15–20.

Steinbock, Bonnie. "Wrongful Life and Procreative Decisions," In *Harming Future Persons: Ethics, Genetics, and the Nonidentity Problem*, edited by Melinda A. Roberts and David T. Wasserman. Dordrecht: Springer, 2009.

Steiner, Hillel. *An Essay on Rights*. Oxford: Blackwell, 1994.

Steiner, Hillel and Peter Vallentyne. "Libertarian Theories of Intergenerational Justice," In *Intergenerational Justice*, edited by Axel Gosseries and Lukas H. Meyer. Oxford: Oxford University Press, 2009.

Stemplowska, Zofia. "Doing More than One's Fair Share," *Critical Review of International Social and Political Philosophy*, 19(5) (2016), pp. 591–608.

Strong, Carson. "Cloning and Infertility," *Cambridge Quarterly of Healthcare Ethics*, 7 (1998), pp. 279–293.

Strong, Carson, Jeffrey R. Gingrich, and William H. Kutteh. "Ethics of Postmortem Sperm Retrieval," *Human Reproduction*, 15(4) (2000), pp. 739–745.

Sumner, L.W. *The Moral Foundations of Rights*. New York: Oxford University Press, 1987.

Taylor, Robert S. "Children as Projects and Persons: A Liberal Antinomy," *Social Theory and Practice*, 35(4) (2009), pp. 555–576.

Thomson, Judith Jarvis. "A Defense of Abortion," *Philosophy and Public Affairs*, 1(1) (1971), pp. 47–66.

Tomlin, Patrick. "Should Kids Pay Their Own Way?" *Political Studies*, 63(3) (2015), pp. 663–678.

Tooley, Michael. "Value, Obligation, and the Asymmetry Question," *Bioethics*, 12(2), pp. 111–124.

UN Department of Social and Economic Affairs (Population Division). *Child Adoption: Trends and Policies*. New York: United Nations, 2009.

UN General Assembly. *Universal Declaration of Human Rights*. 10 December 1948, 217A(III).

UN General Assembly. *International Covenant on Civil and Political Rights*. 23 March 1976, 2200A (XXI).

UN General Assembly. *United Nations International Conference on Population and Development Programme of Action*. 18 October 1994.

UNAIDS, UNICEF, and USAID. *Children on the Brink 2004: A Joint Report of New Orphan Estimates and a Framework for Action*. New York: United Nations Children's Fund 2004.

Unger, Peter. *Living High and Letting Die: Our Illusion of Innocence*. Oxford: Oxford University Press, 1996.

United Kingdom. *Human Fertilization and Embryology Act*, 2008.

Vallentyne, Peter. "Equality and the Duties of Procreators," In *The Moral and Political Status of Children*, edited by David Archard and Colin. M. Macleod. Oxford: Oxford University Press, 2002.

Vallentyne, Peter. "The Rights and Duties of Childrearing," *William and Mary Bill of Rights Journal*, 11(3) (2003), pp. 991–1009.

Van Basshuysen, Phillipe and Eric Brandstedt. "Comment on 'The Climate Mitigation Gap: Education and Government Recommendations Miss the Most Effective Individual Actions,'" *Environmental Research Letters*, 13(4) (2018), pp. 1–3.

Van den Broeke, U., M. Vandermeeren, D. Vanderschueren, P. Enzlin, K. Demyttenaere, and T. D'Hooghe. "A Systematic Review of Sperm Donors: Demographic Characteristics, Attitudes, Motives, and Experiences of the Process of Sperm Donation," *Human Reproduction Update*, 19(1) (2013), pp. 37–51.

Vance, Chad. "Justifying Subsistence Emissions: An Appeal to Causal Impotence," *Journal of Value Inquiry*, 57(3) (2021), pp. 515–532.

Vance, Chad. "Procreation is Immoral on Environmental Grounds," *The Journal of Ethics*, 28(1) (2024), pp. 101–124.

Velleman, J. David. "Family History," *Philosophical Papers*, 34(3) (2005), pp. 357–378.

Velleman, J. David. "Persons in Prospect," *Philosophy and Public Affairs*, 36(3) (2008), 221–288.

Waldron, Jeremy. "A Right to Do Wrong," *Ethics*, 92(1) (1981), pp. 21–39.

Waldron, Jeremy. "Rights in Conflict," *Ethics*, 99(3) (1989), pp. 503–519.

Wasserman, David. "The Non-Identity Problem, Disability, and the Role Morality of Prospective Parents," *Ethics*, 116(1) (2005), pp. 132–152.

Weinberg, Rivka. "The Moral Complexity of Sperm Donation," *Bioethics*, 22(3) (2008), pp. 166–178.

Weinberg, Rivka. *The Risk of a Lifetime: How, When, and Why Procreation May Be Permissible*. Oxford: Oxford University Press, 2015.

Wilkinson, Stephen. *Choosing Tomorrow's Children: The Ethics of Selective Reproduction*. Oxford: Oxford University Press, 2011.

Williams, Bernard. "A Critique of Utilitarianism," In *Utilitarianism: For and Against*, edited by J.J.C. Smart and Bernard Williams. Cambridge: Cambridge University Press, 1973.

Wolf, Susan. "Moral Saints," *Journal of Philosophy*, 79(8) (1982), pp. 419–439.

Woodward, James. "The Non-Identity Problem," *Ethics*, 96(4) (1986), pp. 804–831.

Woolard, Fiona. "Mother Knows Best: Pregnancy, Applied Ethics, and Epistemically Transformative Experiences," *Journal of Applied Philosophy*, 38(1) (2021), pp. 155–171.

Wynes, Seth and Kimberley A. Nicholas. "The Climate Mitigation Gap: Education and Government Recommendations Miss the Most Effective Individual Actions," *Environmental Research Letters* 12 (2017), pp. 1–9.

Young, Thomas. "Procreation and Overconsumption: Are they Morally Equivalent?" *Journal of Applied Philosophy*, 18(2) (2001), pp. 183–192.

Index

accidental fatherhood 75
accidental pregnancy 7, 121, 132–134, 168
Almond, Brenda 191
antenatal depression 86, 98
anti-cost-sharing view 148
anti-natalism 9, 44, 184–185
Archard, David 52, 140
assisted reproductive technology 195, 198–203
asymmetry problem 131–134
Austin, Michael 119–120, 139

baby redistribution problem 26, 77–82, 87, 98
baby scarcity 80, 101
baby surplus 100
Bayles, Michael 54
Benatar, David 58
benefit foisting 156
biological connectedness 7, 29, 76, 83, 97, 118, 134–139, 189, 193
biological parenthood 105, 187–193, 195
bionormativity 202
birthright claim 52
Boonin, David 59–60
Botterell, Andrew 194
Brake, Elizabeth 114–115, 124–128
Brennan, Samantha 30, 78
Brighouse, Harry 24–27, 32, 39–40, 78–82, 85, 100
Brock, Dan W. 23–24, 47
brute luck 81, 151, 158, 175
Bucharest Early Intervention Project 126

Caney, Simon 221–223
Casal, Paula 159–161
causal account of parental obligation 7, 108–110, 112, 114–139
causal responsibility 115
child abandonment problem 114–115, 118, 128
child's right to a decent start 6, 47, 52–54, 60–61, 63–66, 70
Cinderella Effect 136
claim rights 13
cloning 192
collective afterlife 177, 204, 226
compossibility 17
conflicts of rights 16–17
Conly, Sarah 221, 226
consent: hypothetical 59–60; subsequent 60–63
contractualism 165
control condition for moral responsibility 120
core vs. derivative rights 18–19
cost/intentionality objection 159–161
costs of added adult members 152
costs of care 152

Decent Start Principle 48, 68–70
DeGrazia, David 63, 67
delegation view (of parental obligations) 117, 122
demandingness: and the causal account of parental obligation 128–130; and the duty to rescue 183–184; and the interest theory of rights 15–16
desert, and parental rights 84
discrimination, against infertile persons 201–203

duty to rescue 8, 179–189, 199
duty to take up the slack 186

eco-gluttony 209–211
ecological liberalism 221
egoism 184–185
Environmental Argument for Limiting Procreation 208–211, 213, 217–218, 223–225, 228
epistemic condition (for moral responsibility) 121–122
equal per capita view 222
Essential Consumption View 214–218, 224
explanatory problem (of the causal account of parental obligation) 124–128
explanatory problem (of the relationship-based gestational account) 86–87
externalities 8, 154–159, 172–174

Feinberg, Joel 14–15, 56
Ferracioli, Luara 33–35
filial obligation 113
free riding 154, 159–160, 162
Friedrich, Daniel 181–182, 188–189, 195–196

general autonomy principle 95
genetic account of parental obligation 108, 110–113
genetic determinism 192
gestational surrogacy 20–21, 85, 89–92
gestation, burdens of 83–84; significance to the acquisition of parental rights 82–86
Gheaus, Anca 33, 80–95
global justice 173–174
Goldfarb, William 126
goods of childhood 30

Hanna, Jason 122–124
Hannan, Sarah 28–31, 77
Harris, John 18
Hart, H.L.A. 155–156
Haz-Mat theory of parental obligation 112, 146
Heyd, David 54–55
Hohfeld, Wesley 13

identifiability objection 54–55
immortality 192–193
impersonal selection principles 69
inclusivity dilemma 87–90
interest network 14–15
interest theory of rights 4, 12–18, 52, 91
intergenerational Justice 56, 150, 173, 222–224
Intergovernmental Panel on Climate Change 222
investment theory of parental rights 84, 87
in-vitro fertilization 37–39, 110, 119, 123, 195, 199

Klosko, George 156

lack of legitimate complaint objection 63–66
Leland, R.J. 28–31
Levy, Niel 192
Liao, S. Matthew 89, 97
limits of permissible consumption 208, 210–211, 213, 215, 217–228
Locke, John 83
Lotz, Mianna 192

MacIver, Corey 211–214
maternal-fetal bonding 86–87, 94, 97
maximin principle 166–168
McLeod, Carolyn 194
Meijers, Tim 226
Millum, Joseph 84, 111, 134
monism vs. pluralism about parental obligation 110
moral equivalence thesis 209–210, 228–229
moral responsibility, conditions of 119–121
moral responsibility for parental need account of parental obligation 139–140
Murtaugh, Paul A. 207, 212, 216

Noggle, Robert 78
non-comparative harm 50
non-existence objection 56–58
non-identity problem 5, 48–50, 53–54, 68
Nozick, Robert 156

Olsaretti, Serena 126, 148, 152, 160–164, 174
option luck 151, 157, 166–169
orphan crisis 179, 181–183, 187, 198–199, 201
orphan, definition of 178
Overall, Christine 21–22, 37, 221

parental interest justification of the right to procreate 12, 24–41
parental obligation, relationship to parental rights 74–77; theories of 107–110
parental package view 74, 76, 140
parental project view 6, 75, 95–101, 141
parental rights, relationship to parental obligation 74–77
parental role morality 69
parenthood, as a moral role 76–77
Parfit, Derek 48, 54, 58, 61, 64, 69
parity principle 88, 90–91, 99
Pinkert, Felix 216
Porter, Lindsey 86, 122
posthumous reproduction 192
precautionary principle 225, 227
pregnancy, experience of 20–21, 190–191; burdens of 83–84
principle of attribution 152
principle of fair play 155–157, 159, 163–164, 169
principle of procreative beneficence 59
principle of stakes 152
principles of restriction vs. principles of selection 69
procreative costs 124–125
procreative limitarianism 210–228
procreative motives 138
proprietarianism 83, 85
Prusak, Bernard 79, 135–137
public goods 150, 153–163

question of procreative justice 3

Rakowski, Eric 151
rape-related pregnancy 112, 141–142
Rawls, John 155–156, 164–166, 174
Raz, Joseph 13, 18
Reasonably Foreseeable Consumption View 214–217
relationship-based gestational account of parental rights 6, 75, 78, 82–87, 90–98
responsibility-sensitive egalitarianism 152
responsibility thesis 212–217
Richards, Norvin 99
Rieder, Travis 227
rights-based solutions to the non-identity problem 50–54
right to bodily integrity 21–22, 59, 64–65, 134
right to do wrong 139
right to parent 24–29; child-centered accounts of 32–33, 78–79; conditionality of 25, 79; dual interest accounts of 24–29, 32–33, 78–80
right to parent one's biological child 6, 31–33, 75–76, 78, 82, 95–96, 100, 140
right to procreate, direct justifications of 19–21; indirect justifications of 21–24; limits of 36–41; parental interest justification of 24–41, 75
right to self-determination 23–24
Robertson, John 18, 37, 192
role obligations 108, 113–114
Rothman, Barbara Katz 88
Rulli, Tina 179, 181–182, 188–190, 199

Savulescu, Julian 69
Scheffler, Samuel 226
Schlax, Michael G. 207, 212, 216
search movement 137
Shanner, Laura 21–22
Singer, Peter 180, 183–184
Sinnott-Armstrong, Walter 219–220
small family ethic 227
socialized goods objection 161–163
sperm donation 19–21
Steinbock, Bonnie 52
Sticker, Martin 216
strict liability 129–130
Strong, Carson 189
supererogation 101, 125, 170, 188
Swift, Adam 24–27, 32, 39–40, 78–82, 85, 100

Thomson, Judith Jarvis 114
three-parent IVF 110
time-dependence claim 48–49
too many parents problem 118–124
Total Lifetime Consumption View 212–218, 226
transfer view (of parental obligation) 116

unfairness objection 67–68
United Kingdom Human Fertilization and Embryology Act 123–124
United Nations Convention on the Rights of the Child 52
Universal Declaration of Human Rights 11

veil of ignorance 165–168
Velleman, J. David 33, 58, 79, 115–117, 127, 130, 136–138
voluntarist account of parental obligation 7, 108, 110, 113–115
volunteer principle 113

waiving of rights objection 58–63
Waldron, Jeremy 16–17
Wasserman, David 67, 70
Weinberg, Rivka 112
weltanschauung 191–192
Williams, Andrew 159–161
Woodward, James 51, 54
wrongful life 63–64

Young, Thomas 209–211, 215–216

Made in the USA
Las Vegas, NV
19 September 2026

b16d94ff-41cf-4027-9cef-86175d6c2571R01